Building Services, Technology and Design

Roger Greeno

LONGMAN

The CHARTERED
INSTITUTE OF
BUILDING

Pearson Education Limited
Edinburgh Gate, Harlow,
Essex CM20 2JE, England
and associated companies throughout the world

Co-published with The Chartered Institute of Building through
Englemere Limited, The White House
Englemere, Kings Ride, Ascot
Berkshire SL5 8BJ, England

First published 1997
Third impression 2000

British Library Cataloguing in Publication Data
A catalogue entry for this title is available from the British Library

ISBN 0-582-27941-0

Set by 35 in 9/13pt Palatino
Printed in Malaysia, TCP

Contents

1 Cold water supply

Hydrological cycle

The origins of drinking water are found in the hydrological or water cycle. This continuous process includes evaporation of moisture from the sea, rivers and lakes to form clouds. The condensation droplets coalesce and fall as rain to replenish the water levels.

Extraction

In the United Kingdom, water is derived from:

1. Surface sources, i.e. rivers and large lakes
2. Boreholes into aquifers (water-bearing strata)
3. Roofs, paved areas and shallow wells.

The first two categories lend themselves to subsequent filtration and chemical treatment, before distribution and consumption, as shown in Figure 1.1. The latter category could be contaminated and are only used in remote locations.

In some parts of the world where rainwater is scarce, desalination plant is used to process sea water, but this is very expensive.

Processing and distribution

In Figure 1.1 the three principal sources of United Kingdom water are identified. About one-third of consumption comes from each, although regional variations are inevitable, e.g. the Highland areas of Scotland will provide the bulk of water for that area, while the flat Fenlands of East Anglia will need boreholes.

Surface water is stored in man-made reservoirs or impounding reservoirs, the latter created from damming valleys to create catchment areas and to provide the potential for hydroelectricity generation. Coarse and fine sand filtration removes most debris and impurities before a minute amount of chlorine is added to classify the water drinkable. Borehole supplies are well filtered through the strata and normally only require chemical treatment.

After processing, the water is pumped to a high-level storage reservoir or

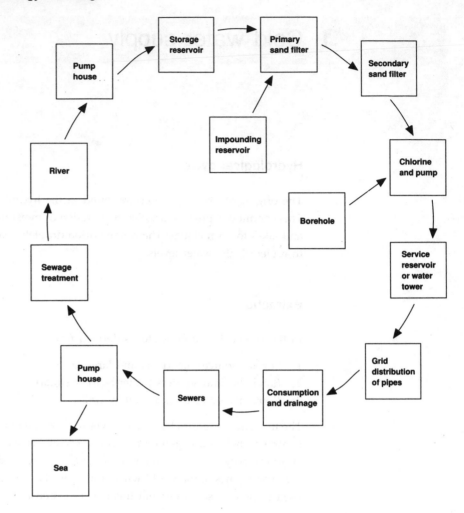

Figure 1.1 The water process

water tower for gravity distribution through iron or uPVC pipes (coloured blue) jointed as shown in Figure 1.2.

Connection to buildings

The grid distribution of underground pipework enables sections to be isolated for repair and maintenance without severe local disruption. Connections to the grid are made by the local water authority or its approved contractor at the expense of the developer or building owner. An isolation stop valve is usually provided at the crown of the water main and a communicating pipe terminated inside the property boundary with another stop valve for the building owner's use. This pipe remains the water authority's, but the service pipe thereafter is the responsibility of the building owner. A typical installation is shown in Figure 1.3.

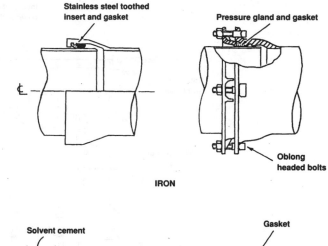

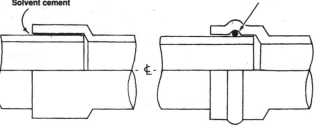

Figure 1.2 Jointing of
water mains

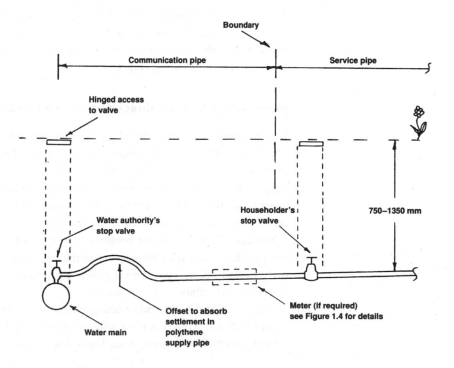

Figure 1.3 Domestic
water supply

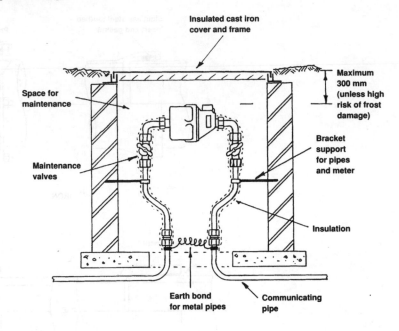

Figure 1.4 Meter housing

Water meters

Water meters are supplied at the discretion of the local water authorities and most new buildings are required to have them. The preferred location is underground, just beyond the property boundary as shown in Figure 1.3. If this is impractical, location within the building at the base of the rising main may be agreed with the water authority. Figure 1.4 shows a possible installation in a waterproof chamber, with an electrical earth continuity bond for metal pipework should the meter be removed. Figure 1.5 is an innovative patent meter connection to an existing stop valve body.

Internal distribution, direct and indirect systems

Direct systems are not favoured by many water authorities, as they require a consistent supply of pressurised water, which may be difficult during periods of peak demand. In the Highland areas this is viable where the sources of supply and distribution are well elevated relative to draw-off points, but for most areas the indirect system is mandatory. Although more expensive to install, with its larger cistern and almost twice as much pipework, it has the advantage of constant water pressure from the storage cistern which reduces the possibility of back siphonage (possible negative mains pressure drawing dirty water back into the main, e.g. hosepipe attached to an outside tap, with the open end submerged in a pond!).

Figures 1.6 and 1.7 provide a comparison of the systems with characteristics of installations. Note that even with an indirect system, the sink tap is connected directly to the rising main. This is the only direct connection permitted

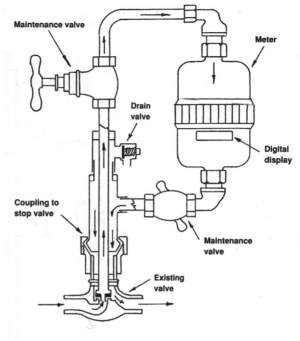

Figure 1.5 Patent meter connection

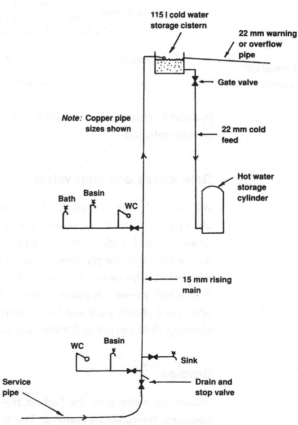

Figure 1.6 Direct cold water supply system

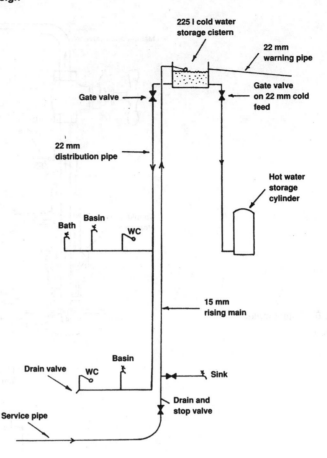

Figure 1.7 Indirect cold
water supply system

in order to ensure a safe drinking water supply, as cistern-stored water could
be contaminated.

Gate valves and stop valves

All mains distribution pipework is subject to high pressure, therefore the
use of gate valves is not recommended in this situation as the metallic gate
or sluice which slides vertically within a guide is likely to vibrate and send
tremors through the pipework. This could cause fractures as well as acoustic
discomfort. Also as the gate wears, it is unlikely to retain the ability totally to
isolate high-pressure supplies, whereas the stop valve contains a plastic washer
which will absorb wear and is easy to replace. Sectional details of both valves
showing their operating features are shown in Figure 1.8.

Storage

Storage of water is in the highest possible location to maintain reasonable
pressures throughout a building. This is normally in the roof space of a house,

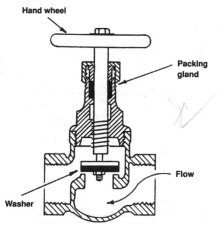

STOP VALVE

Hand wheel

Packing gland

Flow

Washer

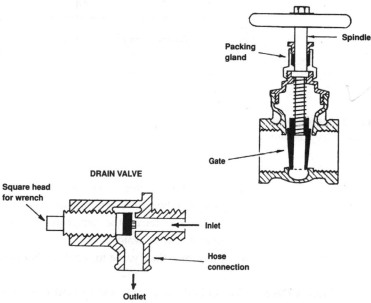

GATE VALVE

Spindle

Packing gland

Gate

DRAIN VALVE

Square head for wrench

Inlet

Hose connection

Outlet

Figure 1.8 Valves

but care must be exercised to ensure that the water is not affected by dust, debris and frost. A typical installation is shown in Figure 1.9, which will provide a reserve if the mains supply is interrupted – it will reduce demand on the main and provide constant pressure to all outlets. The low-pressure supply will also limit wear on taps and fittings, reduce water wastage and noise transmission.

Cistern manufacturers and installers should comply with the following:

- use of non-corrosive, shatter-proof materials
- provide a close-fitting lid with a vent
- adequate insulation
- substantial support with a level platform

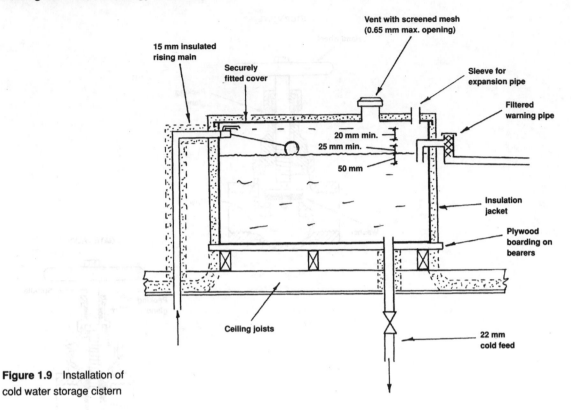

15 mm insulated rising main

Securely fitted cover

Vent with screened mesh (0.65 mm max. opening)

Sleeve for expansion pipe

Filtered warning pipe

20 mm min.
25 mm min.
50 mm

Insulation jacket

Plywood boarding on bearers

Ceiling joists

22 mm cold feed

Figure 1.9 Installation of cold water storage cistern

- fitting of a warning (overflow) pipe larger in diameter than the inlet
- provide valves on every outlet (except warning pipe!)
- on large cisterns, the outlet to be opposite the inlet to encourage throughflow of water
- fitting of a float valve to regulate and control the water level.

Float valves   Float or ball valves are used to control the water level in a cistern automatically. They are manufactured in two distinct types, the traditional piston ball valve with bottom outlet and the now preferred diaphragm valve with a top outlet. Both are shown in sectional detail in Figure 1.10. In order to satisfy current legislation, a substantial air gap is required between cistern water level and float valve outlet, to prevent the possibility of back siphonage if the valve mechanism and overflow fail. This is easily achieved with the top outlet diaphragm valve, but where the bottom outlet piston-type valve is used, a non-return device known as a check valve shown in Figure 1.11 must be applied just before the ball valve. These are also required with outside tap installations.

Cold water storage capacity   Water storage is usually quantified to satisfy a 24-hour interruption of supply. As this is very rare, the designer may choose to reduce the following design guidance accordingly:

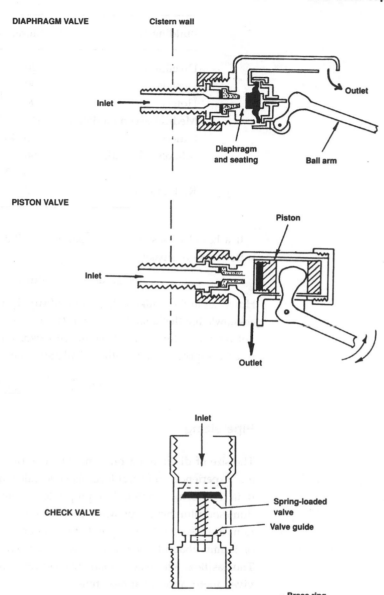

Figure 1.10 Float-operated valves

Figure 1.11 Check valve and copper couplings

Building type	Storage for 24 hours	
Dwellings	90	
Hostels	90	
Hotels	90	
Medical accommodation	115	litres per person
Offices	35–45	
Schools – boarding	90	
day	15–20	
Restaurants	7	litres per meal

If a hostel is designed to accommodate 100 students, determine the cold water storage capacity.

$$100 \text{ students} \times 90 \text{ litres (l)} = \underline{9000 \text{ l}}$$

However, in the unlikely disruption of supply, the designer would be wise to acknowledge that a shut-down for 24 hours is unusual and as the situation is not desperate for water, it would be reasonable to allow perhaps 10 hours' reserve supply. Therefore the calculation could be revised thus:

$$9000 \text{ l} \times \frac{10}{24} = \underline{3750 \text{ l}}$$

Pipe sizing

The size of distribution pipes may be determined by established knowledge and experience, empirical formulae and calculation based on statistical usage data. The former method is appropriate for simple repetitive systems such as domestic plumbing, but when installations become more complex, calculations are essential to incorporate the numerous variables. Formulae and rules of thumb abound. One of the most successful is the formula attributed to Thomas Box, whose aged mathematical calculation of pipe diameter has survived metrication to appear thus:

$$q = \sqrt{\frac{d^5 \times H}{25 \times L \times 10^5}}$$

where q = flow rate (l/s)

d = internal diameter of pipe (mm)

H = head or pressure (m)

L = effective length of pipe (m).

When transposed to make d the subject, Box's formula appears as follows:

$$d = \sqrt[5]{\frac{q^2 \times 25 \times L \times 10^5}{H}}$$

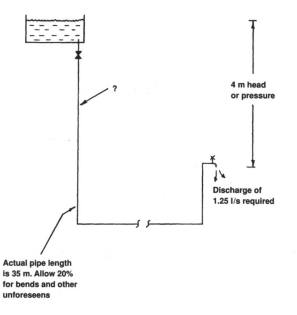

Figure 1.12 Pipe sizing

The simple installation shown in Figure 1.12 provides an opportunity to illustrate an application of this formula:

$$d = \sqrt[5]{\frac{(1.25)^2 \times 25 \times (35 + 20\%) \times 10^5}{4}}$$

$$= \sqrt[5]{410 \times 10^5} = \underline{33.3 \text{ mm}}$$

This is a general pipe-sizing formula applying to all materials, so the nearest commercial pipe size over 33.3 mm would be 38 mm inside diameter steel or 42 mm outside diameter copper. Note, steel pipes are specified internally and copper externally.

Loading unit method   Loading units are the result of statistical surveys and analysis to provide practical guidance based on intermittency of use and an expectation of flow rate at a variety of fitments:

Fitment	Loading units
Wash basin	1.5–3, depending on size
WC cistern	2
Washing machine	3
Dishwasher	3
Shower	3
Sink ($\frac{1}{2}$ in. tap)	3
Sink ($\frac{3}{4}$ in. tap)	5
Bath ($\frac{3}{4}$ in. tap)	10
Bath (1 in. tap)	22

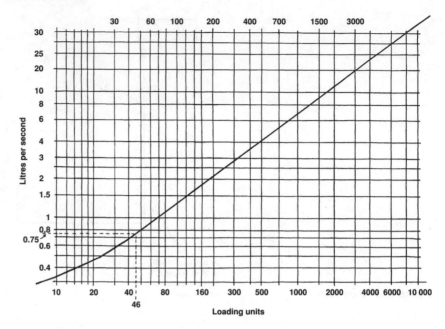

Figure 1.13 Conversion chart

Procedure

- determination of flow rate
- calculation of effective length of pipe run
- calculation of pressure or head loss due to friction.

Determination of flow rate

For industrial machinery and commercial equipment, e.g. vending machines, the desired flow rate can be established from manufacturer's data. For general use and supplies to sanitary fitments, the loading units are summed and converted to a flow rate in litres per second using the chart shown in Figure 1.13.

Calculation of effective pipe length

An allowance for bends, offsets, etc. must be made to the theoretical length of pipework, and turbulence due to valves, tee branches, etc. will also affect the flow conditions. Table 1.1 shows the effect of fittings, but at the design stage it is impossible to ascertain the number of obstructions and deviations a pipeline will have to overcome during installation. Therefore, a sensible increase (generally 10–50 per cent depending on complexity) to the actual pipe length, based on the perceived situation, will normally suffice.

Calculation of pressure or head loss

This is the relationship between potential pressure and the effective length of pipe, found by dividing the former by the latter. This, and the flow rate may be plotted on the nomogram shown in Figure 1.14. The two coordinates are extended to obtain the pipe diameter. This nomogram is specifically for copper

Table 1.1 Effect of fittings on pipe lengths

Internal diameter (mm)	Equivalent pipe length (m)		
	Elbow	Tee	Stop valve
12	0.5	0.6	4.0
20	0.8	1.0	7.0
25	1.0	1.5	10.0
32	1.4	2.0	13.0
40	1.7	2.5	16.0
50	2.3	3.5	22.0

Note: The resistance through a tee is insignificant. The figure given is for the change of direction.

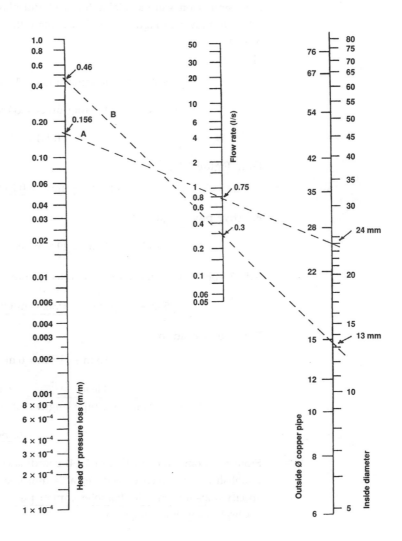

Figure 1.14 Nomogram

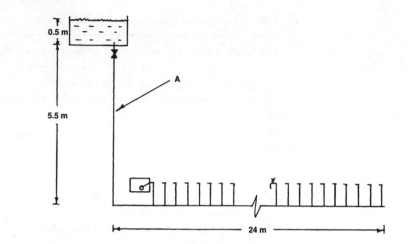

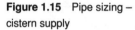

Figure 1.15 Pipe sizing –
cistern supply

tube; similar ones are available for steel and plastic pipes. A practical application is shown in Figure 1.15. To calculate the diameter of the distribution pipe A.

Flow rate:

$$8 \text{ WCs} \times 2 \text{ loading units} = 16 \text{ loading units}$$

$$10 \text{ basins} \times 3 \text{ loading units} = 30 \text{ loading units}$$

$$\text{Total} = 46 \text{ loading units}$$

From Figure 1.13,

$$46 \text{ loading units} = \underline{0.75 \, l/s}$$

Effective pipe length:

$$5.5 \text{ m} + 24 \text{ m} = 29.5 \text{ m actual length}$$

Add 30 per cent to allow for bends, tees, unforeseen deviations, etc.

$$29.5 \text{ m} + 30 \text{ per cent} = \underline{38.35 \text{ m effective length}}$$

Pressure or head loss:

$$0.5 \text{ m} + 5.5 \text{ m} = 6 \text{ m}$$

$$\frac{\text{Head}}{\text{Effective pipe length}} = \frac{6 \text{ m}}{38.35 \text{ m}}$$

$$= \underline{0.156 \text{ m/m}}$$

From the nomogram in Figure 1.14, coordinates of 0.75 l/s and 0.156 m/m establish a line which extends to a 24 mm inside diameter pipe. Therefore specify a <u>28 mm outside diameter copper pipe</u> for this situation.

Check using Box's formula:

$$d = \sqrt[5]{\frac{q^2 \times 25 \times L \times 10^5}{H}}$$

$$= \sqrt[5]{\frac{(0.75)^2 \times 25 \times 38.35 \times 10^5}{6}}$$

$$= \sqrt[5]{8\ 988\ 281}$$

$= \underline{24.6\ \text{mm}}$ inside diameter, i.e. 28 mm outside diameter copper

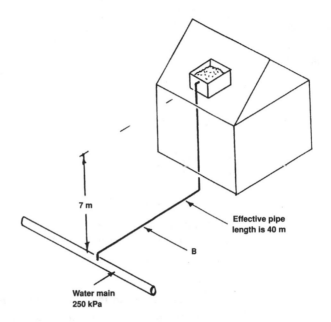

Figure 1.16 Pipe sizing – mains supply

Another application is shown in Figure 1.16, where the diameter of a cistern supply pipe is required from an adjacent water main having a potential of 250 kPa. To calculate the diameter of supply pipe B given a cistern filling rate of 0.3 l/s. Available pressure at the cistern is 250 kPa less 7 m. To convert 250 kPa to metres, divide by gravity, i.e. 9.81. Thus,

$$\frac{250}{9.81} = \underline{25.48\ \text{m}}$$

Therefore the pressure at the cistern is

$$25.48\ \text{m} - 7\ \text{m} = \underline{18.48\ \text{m}}$$

$$\frac{\text{Head}}{\text{Effective pipe length}} = \frac{18.48\ \text{m}}{40\ \text{m}} = \underline{0.46\ \text{m/m}}$$

From Figure 1.14, coordinates of 0.3 l/s and 0.46 m/m extend to show that a 13 mm inside diameter pipe is suitable, therefore specify <u>15 mm outside diameter copper</u>. Check using Box's formula.

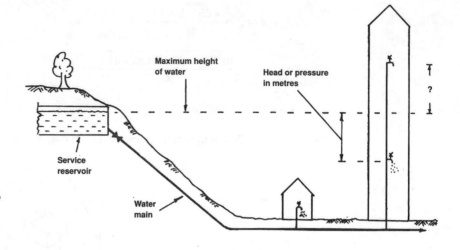

Figure 1.17 Water supply.
Note: Head in metres × gravity = pressure in kPa, e.g. 10 m head × 9.81 = 98.1 kPa

Water supply to high-rise buildings

Many high-rise apartment blocks and offices ascend beyond the natural level of water in adjacent service reservoirs and water towers. Figure 1.17 illustrates this situation and the problem of serving the upper floors in tall buildings.

In these situations, there is a need to boost the water by centrifugal pump to supply the higher levels of tall buildings. Methods vary and consultation with the local water authority is necessary to establish the most acceptable. The basis of system installation can be either:

1. Direct boosting:
   (a) by pumps on the incoming supply pipe (Figure 1.18)
   (b) as (a) but to a storage header (Figure 1.19)
2. Indirect boosting from a break tank:
   (a) to a storage header (Figure 1.20)
   (b) using a pneumatic vessel (Figure 1.21)
   (c) using a continuous running pump (Figure 1.23).

Direct boosting Where mains pressure is relatively high, the water authorities may permit direct boosting, but elsewhere it could impose irregular and uncontrollable demands on mains supply pressure and flow conditions, the worst circumstances being negative pressure or a vacuum and the risk of back siphonage. Where direct boosting is permitted, pump activity is regulated to prevent excessive wear due to frequent and intermittent use by attaching a float switch to the side of the storage cistern. This starts the pump at a predetermined low water level and disconnects it when the water is within about 50 mm of the float valve. Figure 1.18 shows the principle of installation and Figure 1.19 shows an improved variation which provides drinking water from a header vessel. Water for other purposes – bath, WC, etc. – is drawn from the elevated cisterns. The header permits controlled water storage and supply, refilling as

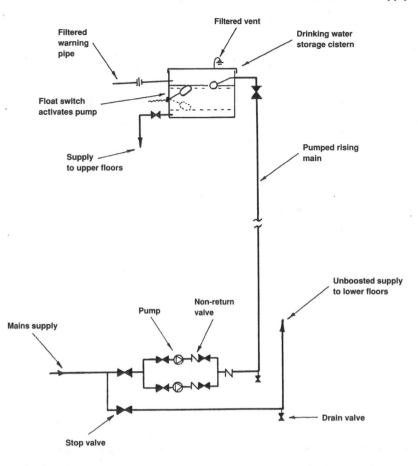

Figure 1.18 Direct pumping to drinking water cistern

the pump replenishes the cisterns. If the header empties before the pump is engaged by the cistern float switch, a pipeline switch detects the low water level in the header and starts the pump. A time delay mechanism or a pressure switch disconnects the pump. Headers are sized at about 5 l capacity per dwelling per day.

Indirect boosting Tank interruption prevents interaction of the boosting pump and mains supply. A typical installation is shown in Figure 1.20 to a drinking water header and cisterns, with pumps combining with a pressure vessel to supply the upper floors in Figure 1.21. The break tank is manufactured from a non-corrosive material such as polypropylene or stainless steel and should be designed and installed to preserve the quality of drinking water. Capacity should equate to approximately 15 min pump output and a low-level float switch is provided to disconnect the pump should the water supply be interrupted.

Use of a pressure or pneumatic vessel is often preferred to the header system, as all the wiring and controls can be contained in a low-level plant room. Air permanently pressurises the water system, until a low-level or pressure

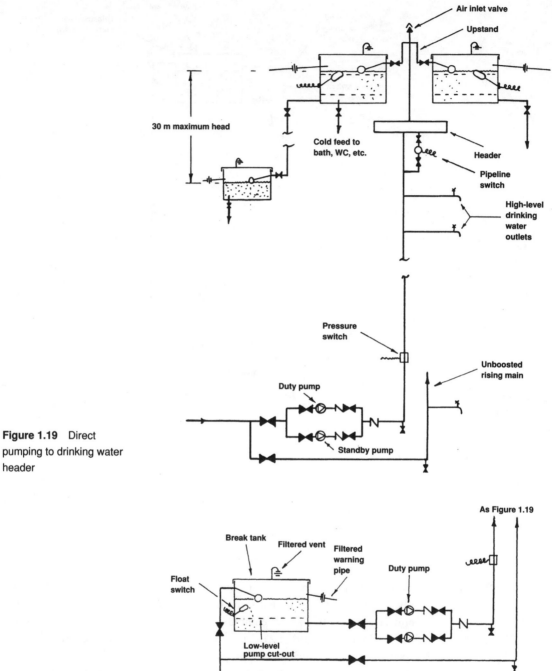

Figure 1.19 Direct pumping to drinking water header

Figure 1.20 Indirect boosting from a break tank

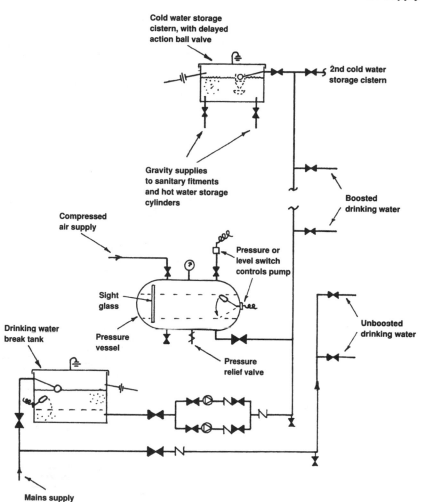

Figure 1.21 Indirect boosting with pneumatic/pressure vessel

switch in the vessel starts the pump to continue supply and to repressurise the vessel. A high-level or pressure switch disconnects the pump and the 'air cushion' re-establishes pressure to supply the upper floors. Air is eventually lost by absorption into the water and can be replenished by air compressor until the correct water level and pressure are recorded in the sight gauge attached to the pressure vessel. Storage cistern float valves supplied from a pneumatic vessel are preferably fitted with a delayed action mechanism to conserve vessel and pump activity. As can be seen in Figure 1.22, the float valve is restrained until a predetermined low water level effects refilling of the cistern.

Continuously running pump systems are popular in modest rise buildings. Installation is relatively simple as shown in Figure 1.23, but care must be taken when sizing the pump. It runs throughout a timed programme, e.g. 07.00–19.00 hours in an office block, and is designed to supply just enough

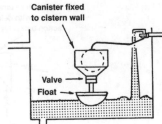

Canister fixed
to cistern wall

Valve

Float

As cistern fills, the
hemispherical float rises
to close the valve

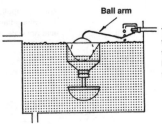

Ball arm

The water level continues
to rise, overflowing into
the canister, lifting the
ball to close off the
water supply

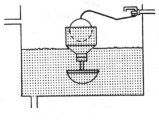

The cistern empties with
demand, but the ball valve
remains closed until the
cistern water level descends,
permitting the hemispherical
float to open the canister
valve releasing the water
retaining the ball float

Figure 1.22 Operation of
a delayed action ball valve

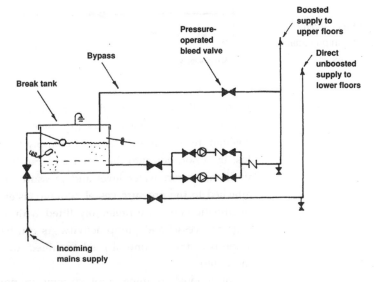

Boosted
supply to
upper floors

Direct
unboosted
supply to
lower floors

Pressure-
operated
bleed valve

Bypass

Break tank

Incoming
mains supply

Figure 1.23 Continuously
running system

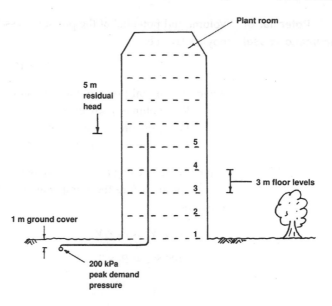

Figure 1.24 High-rise
water supply calculations

water to meet demand. When demand is low, a pressure-controlled motorised
bleed valve opens to allow water to circulate through the break tank.

Boosted water supply
calculations

Figure 1.24 illustrates a typical high-rise situation, with insufficient water
pressure to reach the upper floors. Given that the water authority advises a
minimum mains pressure of 200 kPa, the number of floors that will need a
boosted supply can be calculated thus:

$$\text{Equivalent meter head pressure on the main} = \frac{\text{pressure in kPa}}{\text{gravity}}$$

$$= \frac{200}{9.81} = 20.38 \,\text{m}$$

From this, deduct sufficient residual head as a safety margin of say 5 m, to
ensure a minimal pressure at the top of the rising main and 1 m for the ground
floor level to mains ground cover:

$$20.38 - 5 = 15.38 \,\text{m}$$

$$15.38 - 1 = 14.38 \,\text{m}$$

Divide by floor heights of 3 m:

$$\frac{14.38}{3} = 4.79$$

i.e. five floors unboosted and the remainder up to the plant room boosted.

Potential of pneumatic vessel Volume and potential of the pressure vessel can be assessed by application of Boyle's law, i.e.

$$P_1 \times V_1 = P_2 \times V_2$$

where P_1 = initial pressure (atmospheric – 100 kPa)
V_1 = volume of air in m³
P_2 = pressure of air in use
V_2 = volume of air in use.

For example, using a pressure vessel of 3 m³ total volume with a maximum of 1 m³ of air before the pump interacts, to determine the lowest operating pressure:

$$P_1 \times V_1 = P_2 \times V_2$$

$$100 \times 3 = P_2 \times 1$$

$$P_2 = \frac{100 \times 3}{1} = 300 \text{ kPa}$$

$$\frac{300}{9.81} = 30.47 \text{ m less 5 m residual head} = 25.47 \text{ m potential}$$

At 3 m floor levels,

$$\frac{25.47}{3} = 8.49 \text{ floor spaces}$$

i.e. nine floors, eight plus the roof-top plant room (see Figure 1.24).

2 Hot water supply

Hot water may be provided by individual localised units mounted over sinks, basins, etc. and fuelled by gas burner or an electrical immersion heater. These are a cheap installation for a small cloakroom or similar facility, but it would be uneconomical to provide several of these throughout a house. Therefore, it is usual to provide a centralised heat source in the form of a boiler to heat water in a storage cylinder and to provide sufficient hot water to circulate through several radiators.

Boilers may be floor-standing or wall hung with balanced flue through the wall. Fuel possibilities are conventionally flued solid fuel (coal), gas (mains supply or liquefied petroleum gas (LPG)) or oil. Solid fuel is a popular application for ranges (combined boiler and cooker), but requires regular attention, and therefore has limited automatic control. Gas is very popular and economic where mains supply is possible. It is clean and very adaptable to programmed control and requires little maintenance. Remote situations can also enjoy these facilities with LPG in bottled or delivered supplies. However, these are comparatively expensive. Oil has been the cheapest domestic fuel for many years, but historically has suffered political interference and price fluctuations. Like solid fuel and LPG, oil requires storage space, but is readily adaptable to automatic control.

Centralised systems

Centralised systems have a boiler, a hot water storage cylinder and cold water storage cistern linked by supply and circulatory pipework. Ideally the boiler is close to the hot water storage vessel, to reduce pipework heat losses and to encourage circulation of hot water where gravity or convection is used.

Direct systems　A direct system is shown in principle in Figure 2.1. These are only suited to 'soft' water areas, otherwise the chalk or limestone found in borehole water supplies will precipitate when heated and eventually 'fur up' the boiler and adjacent pipework. As well as rendering the system inefficient, it could lead to a boiler explosion and considerable damage. Furthermore, these systems are not suited to hot water central heating, so are now virtually obsolete in the United Kingdom.

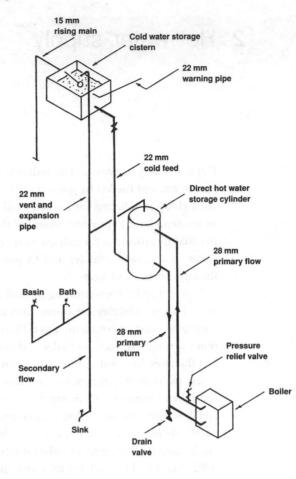

Figure 2.1 Direct hot water supply

Indirect systems With this system an expansion and feed cistern is incorporated to supply the boiler, primary flow and return. The hot water storage cylinder or calorifier shown in Figure 2.2, differs from that in the direct system by incorporating a coil attached to the primary pipework. The resulting system shown in Figure 2.3 enables water to circulate between boiler and coil without being drawn off, therefore the same water circulates persistently, transferring its heat energy into the coil in the surrounding water. Consequently, as no new water is introduced to the boiler, it cannot fur up. Also, water in the calorifier does not mix with the water in the primary flow and return, so radiators can be supplied from these pipes or from separate connections on the same boiler. The expansion cistern merely tops up the boiler and associated pipework should any evaporation occur.

Indirect system with an expansion vessel

As an economy measure, to save installation time, it is possible to incorporate an expansion vessel in the primary pipework, thus eliminating the need for an expansion cistern, expansion pipe and boiler feed pipe as shown in Figure 2.4.

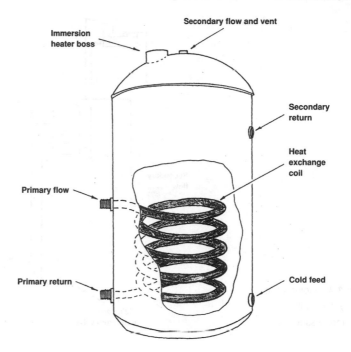

Immersion heater boss

Secondary flow and vent

Secondary return

Heat exchange coil

Primary flow

Primary return

Cold feed

Figure 2.2 Indirect hot
water storage cylinder

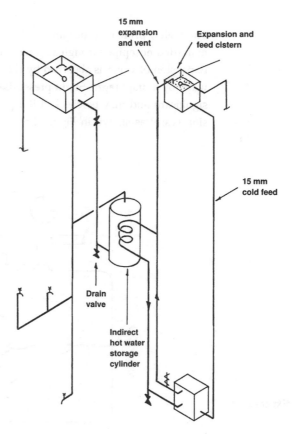

15 mm expansion and vent

Expansion and feed cistern

15 mm cold feed

Drain valve

Indirect hot water storage cylinder

Figure 2.3 Indirect hot
water system

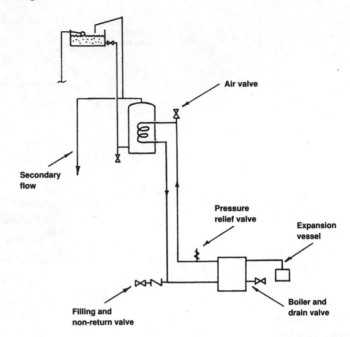

Figure 2.4 Indirect hot water system with expansion vessel

The boiler, primary pipes and coil in the storage cylinder are filled from a mains-fed hosepipe through a filling valve usually located at or near the boiler. Expansion of water is absorbed by the volume of nitrogen separated from the water by a diaphragm. Slight pressurisation may occur, but this should not be significant and may be checked if a pressure gauge is attached to the expansion vessel as shown in Figure 2.5.

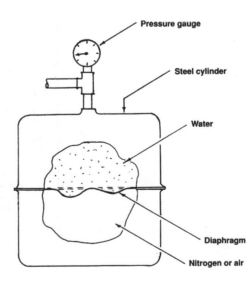

Figure 2.5 Expansion vessel

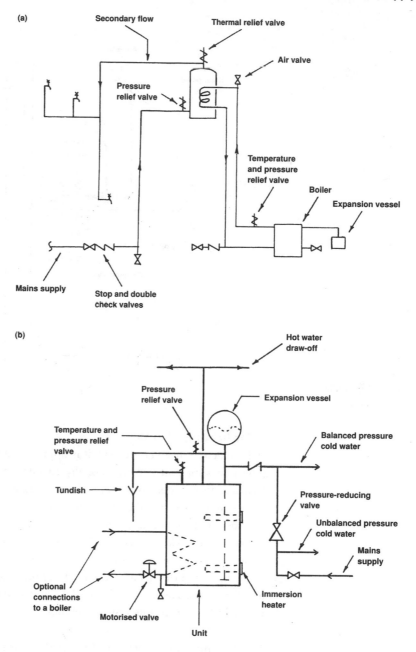

(a)

Secondary flow

Thermal relief valve

Air valve

Pressure relief valve

Temperature and pressure relief valve

Boiler

Expansion vessel

Mains supply

Stop and double check valves

(b)

Hot water draw-off

Pressure relief valve

Expansion vessel

Temperature and pressure relief valve

Balanced pressure cold water

Tundish

Pressure-reducing valve

Unbalanced pressure cold water

Mains supply

Optional connections to a boiler

Motorised valve

Immersion heater

Unit

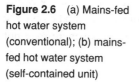

Figure 2.6 (a) Mains-fed hot water system (conventional); (b) mains-fed hot water system (self-contained unit)

Mains-fed indirect hot water system

Mains-fed hot water systems have become a popular alternative to traditional storage supplies. They may only be used in the United Kingdom at the discretion of local water authorities, and must contain both pressure and thermal relief valves for safety. Figure 2.6 shows the application in conjunction with an expansion vessel, thus economising in space as well as installation time.

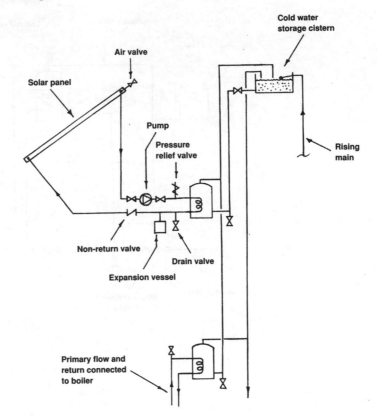

Figure 2.7 Solar hot water supply

Solar heating of water

With 'green' issues very topical, it is appropriate to consider the use of solar power to supplement conventionally fuelled hot water supplies. Solar collectors may be fitted to roofs ideally pitched at about 40° and facing south. They may be linked to the conventional indirect hot water system in the manner shown in Figure 2.7, but space will be needed for an additional heat exchanger. The viability is debatable as the installation expenditure will take several years to recoup, particularly with the unreliable and fickle weather experienced in Britain. Furthermore, the appearance of a house can be severely affected by the panels, and the local authority planning department may not be willing to accept this visual alteration.

Indirect hot water systems for small to medium-sized buildings, e.g. small commercial premises, schools, etc.

Hot water systems for modest-sized buildings are in principle no more than an enlargement of a domestic system. In some cases it may be necessary to specify more than one storage calorifier and a supplementary boiler for stand-by purposes or for the space heating and hot water requirements of the winter months. Figures 2.8 and 2.9 show variations for these larger buildings with secondary circulation to prevent 'dead legs'.

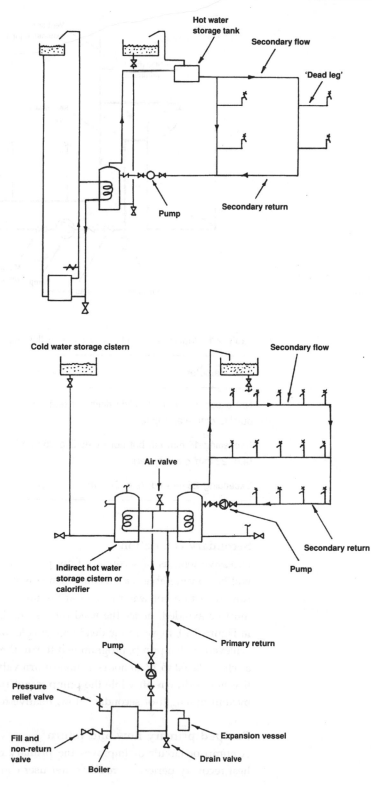

Figure 2.8 Hot water supply to a large house, hotel, school, etc.

Figure 2.9 Hot water supply to commercial premises, e.g. offices

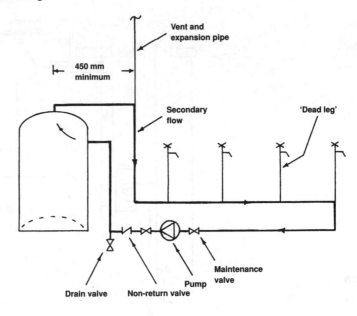

Figure 2.10 Secondary circulation

Table 2.1 Maximum lengths of secondary flow pipes

Pipe diameter	Maximum pipe length (m)
Not exceeding 19 mm inside diameter (e.g. 15 mm outside diameter copper)	12.0
Exceeding 19 mm i.d. but not exceeding 25 mm i.d. (e.g. 22 mm o.d. copper)	7.5
Exceeding 25 mm i.d. (e.g. 28 mm o.d. copper)	3.0

Secondary circulation

Excessive lengths of secondary flow pipe must be avoided, otherwise there will be considerable user inconvenience while waiting for the cold water to run off, before hot water arrives at the taps. Additionally the water wastage must be avoided, hence the need for a secondary flow and return as shown in Figure 2.10, to avoid the dead legs of cold water. Continuous circulation is achieved with a pump, programmed to run throughout the useful day, e.g. in a school, 08.00 to 17.00 hours. A non-return valve prevents reverse circulation if water is drawn off while the pump is not running. Table 2.1 indicates the maximum length of secondary flow, relative to pipe diameter.

Pumped primary flow and return

A pump or circulator improves the primary circulation by accelerating the heat recovery period to create greater user convenience. It can also be useful

Table 2.2 Hot water storage requirements

Building type	Hot water storage (litres per person)
Boarding schools	25
Day schools	5
Dwellings	25–45, depending on type
Factories	5
Hotels	25–45, depending on type
Hostels	35
Hospitals	25–45, depending on type
Offices	5
Sports pavilions	35

where plant location restricts the effectiveness of convected circulation. The expense of a pump is easily justified by the smaller primary flow and return pipe diameter that enhanced circulation permits, and the more rapid heat exchange in the calorifier may allow a reduced storage capacity.

Pipe and plant sizing

This is achieved by processing the calculations in a logical sequence:

1. Storage cylinder (calorifier) capacity
2. Boiler rating
3. Primary flow and return diameter
4. Pump rating.

Table 2.2 provides a list of building types and recommended storage for each occupant.

Example 1   Figure 2.11 shows a typical hot water system for a school designed to serve 70 staff and pupils.

Storage cylinder capacity

$$70 \text{ staff and pupils} \times 5\,l = 350\,l$$

By reference to a cylinder manufacturer's catalogue or British Standard 1566, a 360 l vessel will be selected as the nearest standard size available.

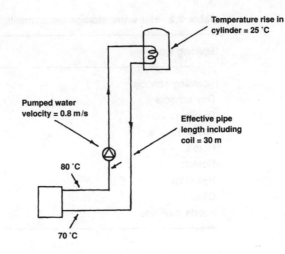

Figure 2.11 Plant sizing
– Example 1

Boiler rating

Boiler power is calculated by applying the formula:

$$kW = \frac{\text{Litres} \times \text{s.h.c.} \times \text{temperature rise}}{\text{Time in seconds}} \times \text{efficiency factor}$$

Interpretation:

s.h.c. = specific heat capacity of water, may be taken as 4.2 kJ/kg K.

Temperature rise = the rise in temperature of existing water in the cylinder to the desired storage temperature of 65 °C. Cold and hot mixing will be about 40 °C so the temperature rise is 65 °C − 40 °C = 25 °C (25 K).

Time in seconds = the amount of time the boiler has to replenish the water through the temperature rise; 1–2 hours is normal.

Efficiency factor = a conventional gas boiler is about 80 per cent efficient, as not all the fuel potential transfers to the water.

Thus:

$$kW = \frac{360 \times 4.2 \times 25}{1.5 \times 3600} \times \frac{100}{80} = 8.75 \ \ kW$$

Primary flow and return diameter

In the interests of efficiency, water circulation in the primary pipes is pumped. Gravity or convected circulation is very slow by comparison, i.e. between about 0.1 and 0.4 m/s depending on plant location and length of pipework. Pumped circulation permits smaller pipe sizes, but as a compromise between economy and efficiency the following should be applied.

Pumped water velocities:

Pipe diameter (internal)	Minimum velocity (m/s)	Maximum velocity (steel) (m/s)	Maximum velocity (copper) (m/s)
Not exceeding 50 mm	0.75	1.5	1.0
Exceeding 50 mm	1.25	3.0	1.5

Note: In the example we are using copper tube with a chosen water velocity of 0.8 m/s.

Boiler rating = 8.75 kW, therefore the actual heat transfer for an 80 per cent efficient boiler is

$$8.75 \times \frac{80}{100} = \underline{7\,kW}$$

Alternatively, the actual heat transfer can be calculated by using this convenient formula:

$$kW = \frac{\text{litres} \times \text{temperature rise}}{\text{time in minutes} \times 14.3}$$

$$= \frac{360 \times 25}{(1.5 \times 60) \times 14.3} = \underline{7\,kW}$$

The mass flow rate in kg/s is

$$\frac{kW}{\text{s.h.c.} \times \text{temperature difference}}$$

Note: Temperature difference is the differential between boiler flow and return. With pumped circulation, water normally leaves the boiler at 80 °C and returns at about 70 °C.

$$kg/s = \frac{7}{4.2 \times (80 - 70)} = \underline{0.16\,kg/s}$$

By referring to the chart in Figure 2.12 the coordinates of 0.16 kg/s and a flow rate of 0.8 m/s indicate that an 18 mm outside diameter copper tube would suffice, but this size is not normally stocked, so <u>22 mm is selected</u>.

Pump rating

Using Figure 2.12, a pressure drop of 350 pascals per metre (Pa/m) of pipe occurs (1 Pa = 1 N/m^2). The pump must overcome this resistance, so

$$350\,Pa \times 30\,m \text{ effective pipe length} = 10\,500\,Pa \text{ or } 10.50\,kPa$$

$$\text{Pump rating} = \underline{0.16\,kg/s \text{ at } 10.50\,kPa}$$

Reference to a pump manufacturer's catalogue shows graphical performance similar to Figure 2.13, where pump No. 2 can be seen as adequately covering the system performance criteria.

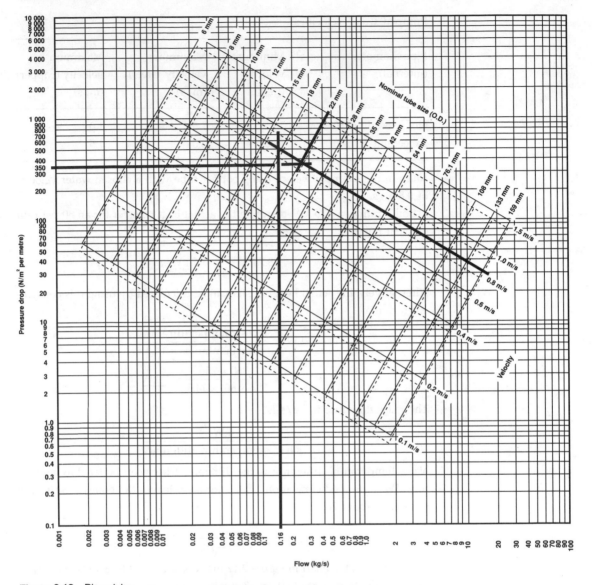

Figure 2.12 Pipe sizing
chart – Example 1:
——————— 65 °C
- - - - - - - - - 115 °C

Example 2 Boiler rating:
(see Figure 2.14)

$$kW = \frac{\text{litres} \times \text{s.h.c. temperature rise}}{\text{time in seconds}} \times \text{efficiency factor}$$

$$= \frac{440 \times 4.2 \times 30}{1 \times 3600} \times \frac{100}{85} = \underline{18\ kW}$$

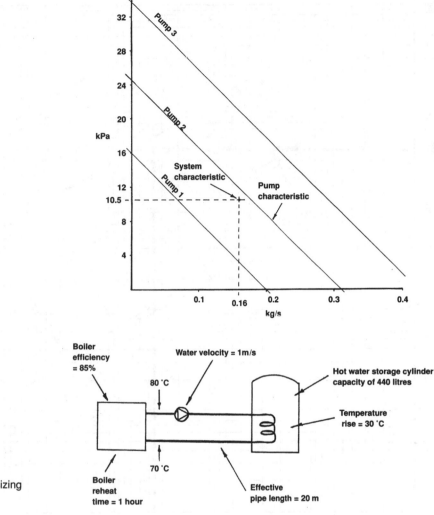

Figure 2.13 Pump performance graph

Figure 2.14 Plant sizing – Example 2

Actual heat transfer to water is

$$18 \text{ kW} \times \frac{85}{100} = \underline{15.3 \text{ kW}}$$

Mass flow rate is

$$\frac{\text{kW}}{\text{s.h.c.} \times \text{temperature difference}} = \frac{15.3}{4.2 \times (80 - 70)} = \underline{0.37 \text{ kg/s}}$$

From Figure 2.15 the coordinates of 0.37 kg/s and pumped water velocity of 1 m/s fall between 22 mm and 28 mm outside diameter copper tube, therefore select 28 mm.

Using a pressure drop of 380 Pa/m, the pump rating is

$$380 \text{ Pa} \times 20 \text{ m} = 7600 \text{ Pa} \quad \text{or} \quad 7.6 \text{ kPa, i.e. } \underline{0.37 \text{ kg/s at 7.6 kPa}}$$

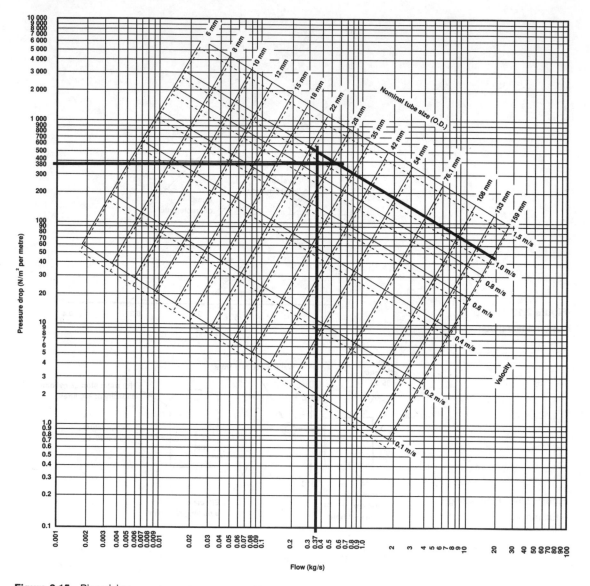

Figure 2.15 Pipe sizing chart – Example 2:
——————— 65 °C
---------- 115 °C

Hot water supply to very large buildings

To provide an efficient distribution of hot water over the large distances experienced in industrial and some commercial developments, it is more economical to pressurise water in a sealed system. The effect is to permit water to reach

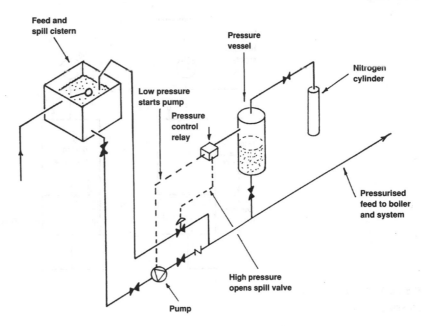

Figure 2.16 Hot water system pressurisation equipment

temperatures exceeding the normal boiling point of 100 °C, without converting to steam. Most systems operate at 120–135 °C, between 300 and 400 kPa gauge pressure (add 1 bar or 100 kPa for absolute pressure). Temperatures up to 180 °C are possible, with pressures exceeding 1 MPa. It is imperative that the pressure increases in line with temperature, otherwise the water will 'flash' into steam and malfunction, possibly damaging plant and equipment.

The most popular method is to pressurise with nitrogen, as this is an inexpensive clean gas, less corrosive than air and less soluble in water. Figure 2.16 shows the installation principle.

Commissioning   Initially the system is charged with water by the duty pump, from water in the feed and spill cistern. The pressure vessel is allowed to fill to one-third of its capacity at atmospheric pressure, when the pump ceases. This is zero gauge pressure as shown in Figure 2.17. The nitrogen is introduced at half design working pressure, which gradually doubles as the boiler increases the water to design temperature. The effect is illustrated in Figure 2.17, with a carefully designed pressure vessel to accommodate the expansion.

The supply cistern, in addition to having a feed function, is used to accommodate excess pressure, when a high-pressure switch opens a motorised spill valve to discharge the surplus water. This is a safety feature which could function if the thermostatic boiler controls fail. A low-pressure switch operates during cool water conditions or when a leak occurs in the system. This engages the duty pump to re-establish correct water and pressure levels.

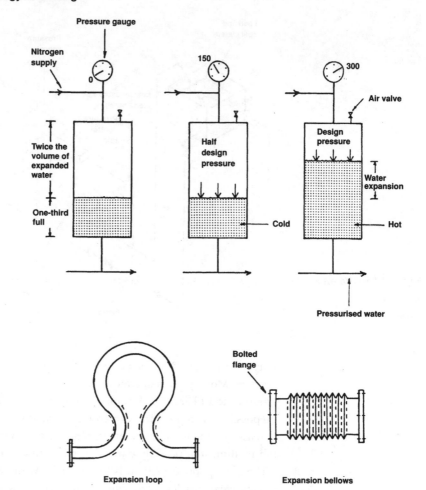

Figure 2.17 Operation of pressure vessel

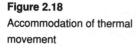

Figure 2.18
Accommodation of thermal movement

Material and safety considerations

As water temperatures and pressures are very high, it is essential that heavy grade steel pipes and equipment is specified throughout. As a safety as well as an economy measure, all pipework must be insulated.

Thermal movement

The amount of thermal movement is considerable, to the extent that expansion bellows or loops as shown in Figure 2.18, must be included at strategic intervals. For example, if a 30 m length of steel pipe (coefficient of lineal expansion = 0.000 012 m °C) is to accommodate a temperature range from 10 to 130 °C, its movement will be

$$30 \text{ m} \times (130 \text{ °C} - 10 \text{ °C}) \times 0.000\,012 = 0.0432 \text{ m or } 43.2 \text{ mm}$$

3 Thermal systems and distribution

Fuels

Practical considerations

Selection of the most appropriate fuel is likely to depend on:

- the personal preference of the system designer or the end user
- pressure from sales promotions
- installation or capital cost of the system
- aftercare, maintenance and replacement costs
- restrictions, i.e. space for fuel storage and exhaust emission legislation
- adaptability to automatic control
- current and projected fuel costs.

Solid fuel

Traditionally, logs of timber and coal products have provided the source of heat energy from open fires and stoves. Both have the disadvantage of considerable space requirement for storage and the need for regular attention to the appliances. The exception are some of the highly efficient anthracites, which lend themselves to hopper-fed supplies and several days' burning potential, as shown in Figure 3.1. There is still the need for ash and dust removal, which

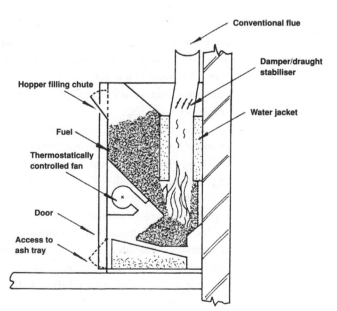

Figure 3.1 Gravity-fed solid fuel boiler

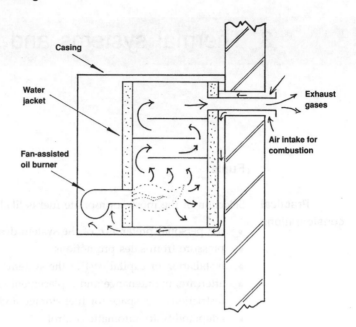

Casing

Water jacket

Fan-assisted oil burner

Exhaust gases

Air intake for combustion

Figure 3.2 Room sealed, balanced flue oil-fired boiler

can be offset against the visual attraction and impact of modern stove and open fire design. Flue and chimney arrangements are larger and more complicated than for other fuels (Building Regulations, Approved Document J) and legislation relating to smoke emission may limit their use. Furthermore, thermostatic control has limited effect; unlike oil or gas fuels, coal burners/boilers cannot be instantly shut off and reignited. However, as a supplementary heat source, they have much to commend them.

Oil

Oil as a boiler fuel is produced by distilling crude refined oil. Various grades are achieved, but for low-level flue discharge, where the percentage of noxious gases must be minimised, the Building Regulations permit only classification C2. This is commonly known as kerosene and is used in vaporising or atomising burners. It is spark ignited and diluted with a fan delivery of air as shown in Figure 3.2.

Oil requires space for storage in a tank. This must not be closer than 1.8 m from the building wall, unless a fire-resistant barrier is sited between tank and building. Figure 3.3 shows the installation and controls necessary for supplying fuel oil to a boiler. Flue systems are less complex than for solid fuel, although a lined chimney may be utilised as well as patent prefabricated flue pipe systems. Horizontal flue discharges are also allowed, but generally only for the smaller domestic units. Maintenance is minimal and an annual service is normally adequate, requiring slightly more attention than an equivalent gas-fired appliance. Thermostatic control of heat output to radiators and hot water is prompt and well defined within the bounds of a time-controlled programmer.

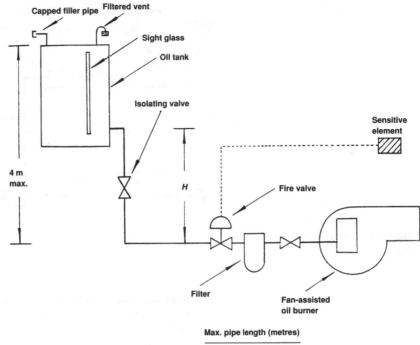

Figure 3.3 Gravity-fed oil supply

H = Head (m)	Max. pipe length (metres)	
	8 mm bore	10 mm bore
0.5	10	20
1.0	20	40
1.5	40	80
2.0	60	100

Gas As a fuel for combustion in hot water appliances, gases are varied in composition and properties. From the Industrial Revolution to the early 1970s, UK consumers used town gas produced from coal carbonisation. This was distributed through a network of underground mains from production plant to end user. Here it combined as a fuel for boilers and an energy source for lighting before the general availability of electricity in the early part of the twentieth century.

Exploration and development of North Sea resources have provided a plentiful supply of natural gas to supersede town gas for mains distribution throughout the UK. It is cleaner to use and has a calorific value, approximately double that of town gas.

Liquefied petroleum gas (LPG) is a convenient alternative to natural gas where mains distribution is unavailable. It is marketed as butane and propane, both readily liquefied and stored in containers under moderate pressure at atmospheric temperatures. Typical storage and supply arrangements are shown in Figure 3.4. LPGs are heavier than air, therefore leakage can be particularly dangerous as the vapour will collect at low level. Appliances cannot be interchanged with natural gas supplies, as the calorific value of the gases

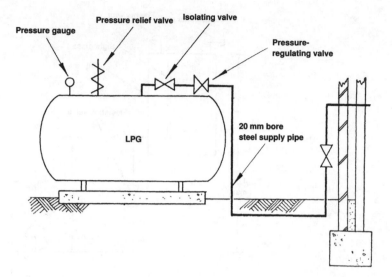

Figure 3.4 Containerised gas supply

Table 3.1 Approximate calorific value of fuels

Fuel	Calorific value
Wood	19 MJ/kg
Coal	30 MJ/kg
Anthracite	32 MJ/kg
Coke	27 MJ/kg
Kerosene	46 MJ/kg
Fuel oil	42 MJ/kg
Town gas	20 MJ/m³
Natural gas	36 MJ/m³
LPG	49 MJ/kg
Electricity	3.6 MJ/KWh

is considerably different as shown in Table 3.1. Equipment manufacturers will advise on the interchangeability of burners.

All gas-burning appliances require little maintenance and respond promptly to thermostatic controls. Flue systems are simple and natural gas benefits from mains supply, eliminating the need for storage.

Economic considerations

Prior to the 1960s, choice of fuel was limited by availability and means of combustion. Solid fuel deliveries were the order of the day and frequent attention boilers/fires the means of combustion. Mains gas was largely restricted to towns and cities, and developments in small-scale domestic gas and oil boilers were at a very early stage. Central heating and its associated controls were restricted by the relatively bulky technology of the time. Few householders could accommodate or afford a centralised hot water and heating system. In the latter part

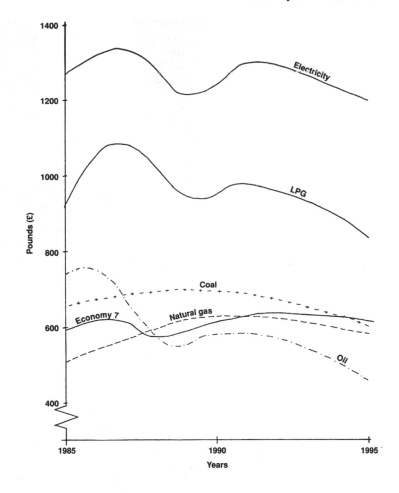

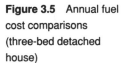

Figure 3.5 Annual fuel cost comparisons (three-bed detached house)

of the 1960s, research and development coupled with expectations of a higher standard of living (Parker-Morris Report, 1963), led to major advances in heating technology. Gas vied with oil as the cheapest and most convenient fuel, oil losing out in the 1970s as political disputes in the source countries of the Middle East escalated the price.

It was not until early in the 1980s when North Sea oil resources became available, that Britain became a credible oil-producing nation and could influence the price of crude oil. Thereafter, domestic fuel oil prices once again compared with mains gas. Figure 3.5 shows a graphical comparison of all available fuels with annual costs typical of a modern three-bedroom detached house requiring hot water and heating in the UK. These approximate values do not take into account regional variations. Declining premiums are the result of price competition between the fuel producers as well as advances and improvements in insulation standards and heat-producing appliance efficiency. (*Note:* Economy 7 relates to off-peak electricity, supplied between 23.00 and 06.00 hours.)

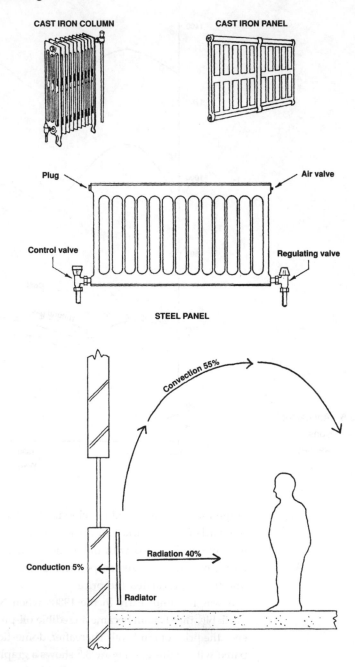

Figure 3.6 Radiators

Figure 3.7 Heat transfer from a radiator

Heat emitters

Various forms of heat emitters are available to suit different situations.

Radiators
These are the most common, with some variations shown in Figure 3.6. The modern format is pressed steel corrugated panelling with finned backing to increase the convective effect. Figure 3.7 shows the pattern of heat transfer,

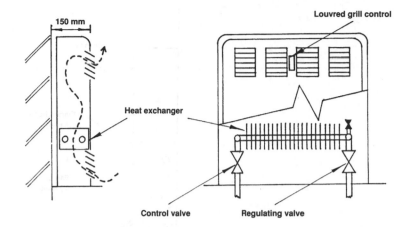

Figure 3.8 Convector

from which it can be observed that most of the energy is convected. This fig-ure increases significantly with the arrangement of finning attached to the back of the emitter. Also, a well-insulated wall and a sheet of reflective paper behind the radiator will reduce unwanted losses in conduction and radiation, respectively.

Convectors Convectors consist of a metal casing and finned pipe heating element as shown in Figure 3.8. The narrow casing and low-level element combine to generate a 'stack effect', where the warm air loses density and gains in velocity in the narrow confines of the casing, to discharge 90 per cent of the heat energy by convection at a modest velocity. These units are comparatively large, being about 1 m high and wide, projecting about 200 mm from the wall; therefore they are most appropriate in offices, entrance halls, classrooms and similar size rooms. To deliver warm air over larger areas, fan-assisted convectors may be selected, or possibly the full perimeter skirting variation. Skirting units are not as efficient, but they do provide well-distributed heat. Figure 3.9 illus-trates functional principles of both.

Radiant panels These consist of a flat panel with looped or coiled pipework attached to the back, as detailed in Figure 3.10. They can be arranged flush with the wall or ceiling to retain continuity of finish, but they are most frequently suspended from workshop or warehouse roof frames to allow clear space around the walls. The back of the panel must be very well insulated to concentrate about 65 per cent radiated heat output from the face.

Exposed pipes Exposed pipework is an economic means of heat provision, but not the most effective or attractive. It is suitable in workshops, factories and other situ-ations where appearance is unimportant. Figure 3.11 indicates a simple triple loop configuration of 80 or 100 mm diameter pipes and Table 3.2 gives the expected heat emission from various pipe sizes.

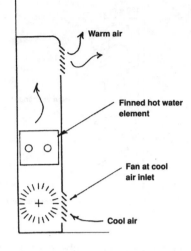

FAN CONVECTOR

Warm air

Finned hot water
element

Fan at cool
air inlet

Cool air

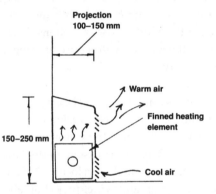

Projection
100–150 mm

Warm air

Finned heating
element

150–250 mm

Cool air

SKIRTING CONVECTOR

Figure 3.9 Convector
variations

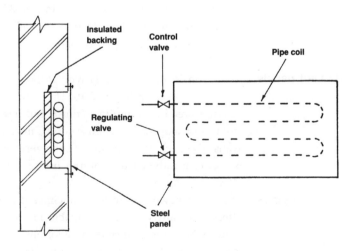

Insulated
backing

Control
valve

Pipe coil

Regulating
valve

Steel
panel

Figure 3.10 Radiant
panel

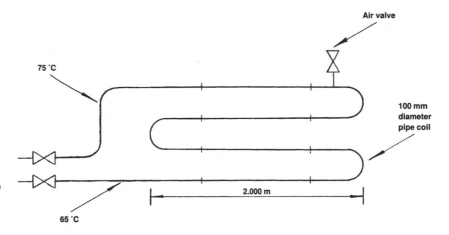

Figure 3.11 Pipe coil.
From Table 3.2, emission
for 100 mm steel pipe at
50 °C temperature
difference (pipe average to
air) = 2 m × 4 pipes ×
230 W = 1840 W

Table 3.2 Approximate heat emission in W/m from steel pipes for ambient air at 20 °C

Pipe diameter		Average water temperature		
In.	mm	60 °C	70 °C	80 °C
2	50	100	130	170
$2\frac{1}{2}$	65	120	160	200
3	80	140	180	230
4	100	170	230	290
6	150	230	320	400

Overhead unit heaters   These units consist of a bank of finned hot water pipes in a casing, with a fan to provide air direction and velocity. They are suspended from the ceiling or roof frame as shown in Figure 3.12 and provide a useful heat source for workshops, warehouses, retail centres and sports halls. Adjustable louvres direct the warm air and in the summer months with the heating off, the fan can be used solely to generate air movement.

Embedded panels   These provide invisible or fabric heating, as the emitter is unseen, contained within the fabric of construction. While offering greater flexibility in room layout and planning, the disadvantage is that the copper, steel or polypropylene pipes conveying hot water, take some time to dissipate their heat through the fabric. Thus, thermostatic response will be slow. Pipework is equally well contained within a wall or ceiling. In the interests of comfort and heat output, the following surface temperatures are recommended:

Ceiling	49 °C
Wall	43 °C
Floor	29 °C

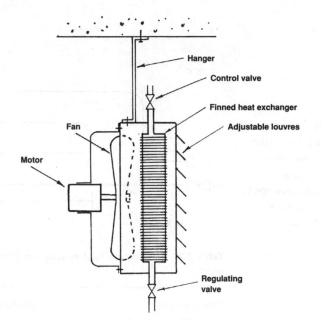

Figure 3.12 Overhead
unit heater

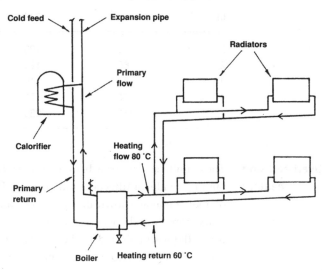

Figure 3.13 Gravity hot
water and heating system

Low-pressure hot water heating systems

These are applied to domestic and other small-scale situations. They operate
at atmospheric pressure with water temperature maximising at about 80 °C.
Before the introduction of small-scale high-efficiency controls, circulation of
hot water from boiler to radiators responded to gravitation or convection currents. This necessitated careful location of emitters, above and not too far from
the boiler. Additionally, fairly large pipes were needed to provide sufficient
mass of water at varied temperatures to create density differentials and gravita-
tional flow. Figure 3.13 shows the principle, which still has an application in
remote areas, where there is no electricity and solid fuel is the only viable fuel.

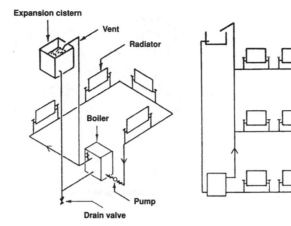

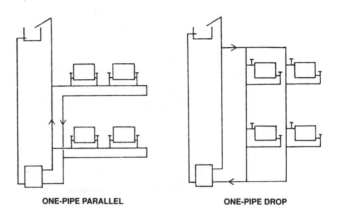

Figure 3.14 One-pipe
heating systems

Contemporary systems have evolved from research in the 1950s and 1960s,
to create small but effective circulating pumps, use of small-diameter pipes,
high-efficiency boilers and responsive controls. Positioning of radiators relative to boiler height is unimportant (within reason), as is apparent from modern wall-mounted boilers above the level of ground floor radiators.

One-pipe systems   Distribution pipework could be one pipe, of the variations in Figure 3.14, but
these economic installations have limited effectiveness. They are difficult to
balance, as controlling one radiator disturbs the flow conditions to others, and
water losing heat in the radiators flows back into the same pipe serving subsequent radiators, thus making the index radiators (those at the end of the
system), hard to heat. Also, if the pump velocity is increased too much, most
of the hot water will bypass the emitters.

Two-pipe systems   These have greater capital outlay, but more effective distribution of water and
control of radiators justify this method for modern practice. There are various

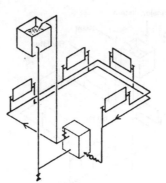

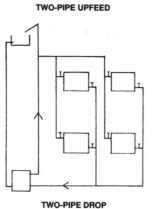

TWO-PIPE REVERSE RETURN   TWO-PIPE UPFEED

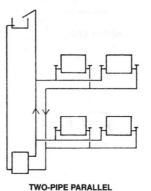

Figure 3.15 Two-pipe
heating systems

TWO-PIPE PARALLEL   TWO-PIPE DROP

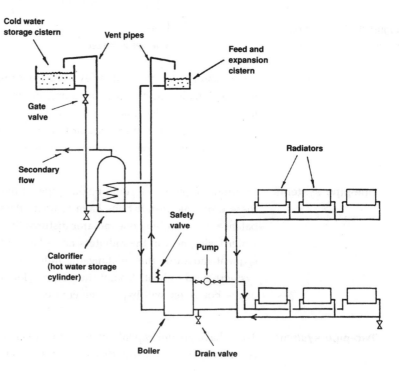

Figure 3.16 Traditional
domestic small-bore hot
water and heating system

configurations of pipe distribution to suit location of service ducts and voids, some of which are shown in Figure 3.15. Figure 3.16 is a schematic combined heating and hot water system, typically applied to modern housing.

Pumped primary and heating circuits

The traditional convected/gravity circulation between boiler and calorifier is slow, uneconomic in fuel consumption and excessive in pipework. The less costly variations in Figures 3.17 and 3.18 have primary flow and return pipes combining with heating flow and return, to provide pumped hot water to calorifier and radiators. The diagram in Figure 3.17 has a cylinder thermostat controlling a motorised valve on the primary return and a separate motor-ised valve controlled by a room thermostat on the heating flow, thus effect-ing independence of the two systems. Figure 3.18 is slightly more advanced,

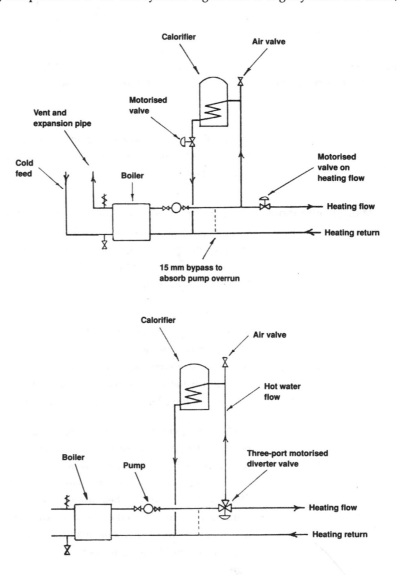

Figure 3.17 Pumped primaries and two-valve hot water and heating system

Figure 3.18 Pumped primaries with three-port diverter valve

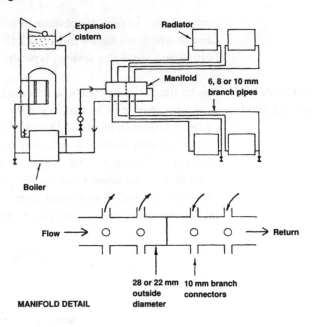

Figure 3.19 Mini/
microbore heating system

using a three-port motorised diverter valve. Subject to thermostatic demands, the pumped water can serve both heating and stored hot water, or either independently.

Mini or microbore The term 'small bore' was derived from the reduced diameters of pipes possible in pumped circulatory systems. Mini or microbore is a stage further, using relatively tiny pipe sizes that were previously limited to refrigeration, gas and engineering applications. Pipe systems of 6, 8 or 10 mm diameter have many advantages:

- simple and quick installation
- less structural disruption
- neater appearance
- low water content, therefore fuel economies
- rapid water circulation, therefore quick thermal response
- pipe in long lengths of soft copper coils, so less jointing.

The very small diameters may be more prone to blockage, although it could be argued that faster water circulation will prevent this. Nevertheless, the high water velocity generates more noise than equivalent small-bore systems. Standard pumped 22 or 28 mm copper flow and return pipes supply strategically located manifolds. These are divided into flow and return compartments, with micro-branches to the various radiators. Figure 3.19 indicates a possible arrangement, with only one manifold shown, but in practice it is usual to apply a manifold for each floor.

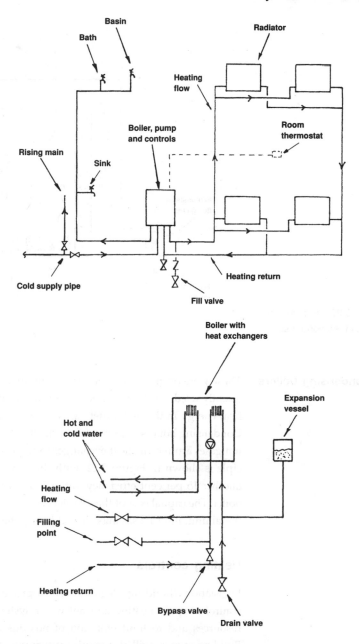

Figure 3.20 Hot water and heating supply from a combination boiler

Figure 3.21 Combination boiler – schematic diagram of controls

Combination boilers

Combination boilers are generally wall mounted, gas- or oil-fired appliances, designed to provide instantaneous hot water for draw-off points and to radiators. Boilers are an integral unit, pre-plumbed and pre-wired for simple compact installation. The principal advantage is the elimination of a hot water storage cylinder (calorifier), as the boiler acts as a mains-fed multi-point supplying hot water to bath, basin, sink, etc. Hot water to radiators is pumped through a heat exchanger and may be conventionally open-vented to a feed and expansion cistern, or more conveniently sealed as shown in Figures 3.20 and 3.21.

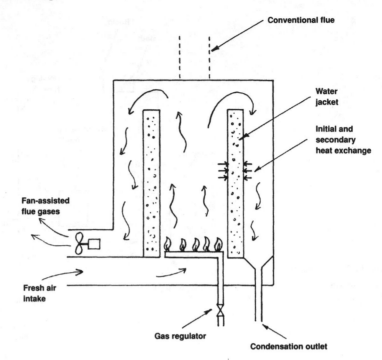

Figure 3.22 Principle of the gas-fired condensing boiler

Condensing boilers These were originally developed for commercial applications, but are now frequently used for domestic hot water and heating. In conventional boilers the heat energy in the burnt fuel gases is lost to the atmosphere through the flue. Condensing boilers deploy a fan in the flue to recirculate the otherwise wasted hot gases around the heat exchanger to create a double heating effect. The principle is shown in Figure 3.22, with the secondary exchange providing about another 15 per cent efficiency to the 75 per cent expected from a simple flued boiler. The capital cost of these boilers is approximately double that of a conventional unit, but fuel savings should justify the expenditure within a few years.

Heating controls

Legislation (Building Regulations, Approved Document L) governing the control of space heating and hot water systems, is necessary to preserve fuel resources and to limit emission of noxious fuel gases into the atmosphere. Building owners will also require automatic controls to regulate fuel consumption and to optimise room temperatures throughout the day. Systems and components range in sophistication from a simple time clock to computerised energy management systems. Whatever is specified, the objective is to control the following:

- temperature of water in the system
- operating hours
- temperature within the space to be heated.

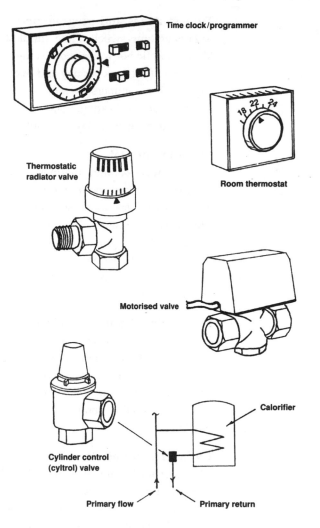

Figure 3.23 Heating and hot water controls

This is to economise in fuel and to maintain comfort and convenience for the building occupants.

Control types
(see also Figure 3.23)

Boiler thermostat

These are in two categories:

1. **Working thermostat**. This is an integral part of the boiler and permits manual control of the water temperature in and leaving the boiler. A control knob, usually numerically scaled from 1 to 5 or 6, engages or disengages the fuel supply at approximately 65–85 °C water temperature.
2. **High limit thermostat**. This too, is an integral part of the boiler and is normally preset by the manufacturer at 90 °C. It is primarily a safety device to close the fuel supply if the working thermostat fails.

Time switches and programmers

A 24-hour time switch is the simplest means of automatic system control. Most domestic hot water and heating systems are adequately controlled with one or two functions per day and override facilities to bypass the clock when required. Programmers are more sophisticated, allowing independent control of hot water and heating. Some can be set to provide 7-day or 28-day functions in advance, with day omit options and several on/off functions per day.

Room thermostat

This is a wall-mounted thermostatically controlled switch, set manually to achieve the desired room or zone temperature. Within the unit, a simple bimetallic strip responds to surrounding air temperature. For most residential situations, a setting of about 20 °C is adequate. Below this, the switch will engage the boiler, pump and possibly motorised valves to direct hot water to the emitters until the preset temperature is attained. Location should be about 1.5 m above floor level, away from draughts and in a central position for the area being heated.

Frost or low limit thermostat

These are positioned on the external north face of buildings. They provide a safety facility for buildings left unoccupied or for unexpected temperature reductions, particularly overnight. They operate independently of time clocks and programmers and will engage the boiler and heating pump when outside temperatures fall to about 2 °C, thus preventing frozen plumbing within the building.

Cylinder thermostats

Hot water for heating averages about 75 °C, while stored hot water for bath, basin, etc. is normally about 60–65 °C. Consequently, separate thermostatic controls are necessary for economic use of the boiler. Also, excessively hot water in the calorifier could cause scalding at the draw-offs. Two types of thermostat exist:

1. **Electrical**. A thermostat is strapped to the side of the calorifier to operate a motorised valve on the primary return and possibly the pump in accelerated systems.
2. **Mechanical**. Independent of electricity, these devices contain expanding and contracting bellows which respond to temperature change. They are fitted to the primary return and when stored water temperature is achieved, the bellows expand to prevent unnecessary circulation. They are only appropriate for gravity/convection circulation.

Thermostatic radiator valves

These are mechanical valves with a built-in sensor which responds to the air temperature surrounding it. Changes cause a wax, liquid or vapour within the valve to expand or contract, regulating hot water flow through the attached

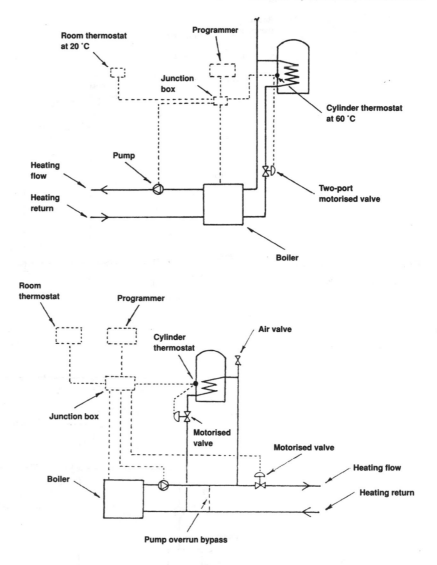

Figure 3.24 Partial control of hot water

Figure 3.25 Two-valve system

radiator. Several settings are provided and these will automatically control room temperatures with variable use, i.e. where occupancy fluctuates or where another heat source such as an open fire is occasionally used.

Motorised valves

These are electrically operated valves which may be used to provide independent temperature control in heating and hot water systems. They are manufactured with two or three ports to suit the different installation plans shown in Figures 3.24–3.26. In larger installations, several room thermostats are deployed throughout the building, each connected to a motorised valve and subcircuit for area or zone control. This is particularly useful where southerly exposed parts of a building are subject to solar gains, therefore requiring less artificial heat input.

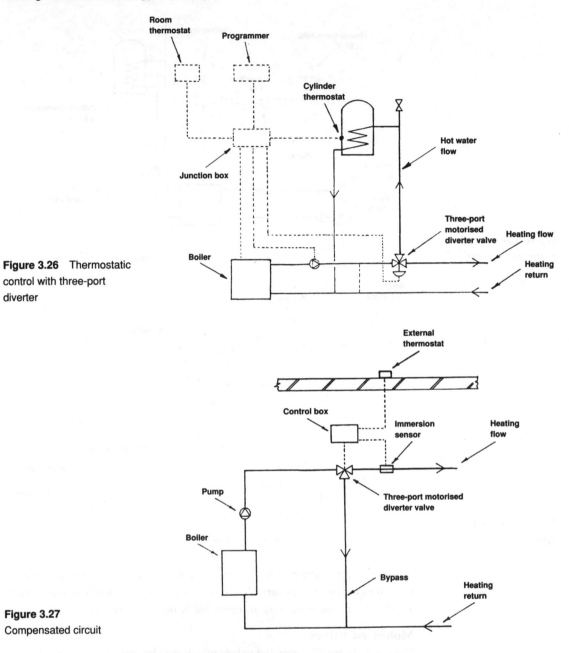

Figure 3.26 Thermostatic control with three-port diverter

Figure 3.27 Compensated circuit

Compensated circuit

A compensated circuit varies the water temperature in a heating system, relative to external temperature. The warmer the outside air, the cooler the water in the system. Figure 3.27 shows the principle, with external thermostat and immersion thermostat effecting position of the three-port motorised valve. This application makes for economic use of heating in medium-sized buildings, but is limited in flexibility on a large scale.

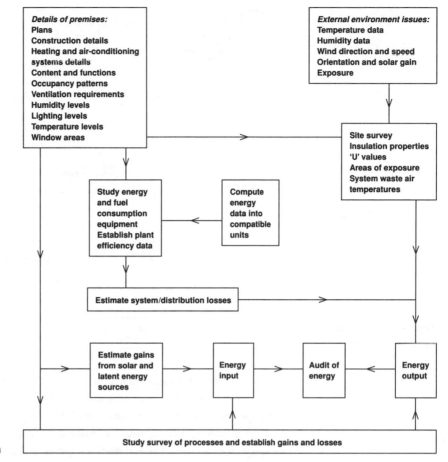

Figure 3.28 Energy
survey/audit flow diagram

Optimum start controller

Temperature varies from day to day and throughout the heating season. It is therefore unnecessary for the boiler to engage at the same programmed time every morning, as on many occasions a later commencement would be sufficient to bring the building up to temperature before occupancy. An optimum start controller is basically a computer, which balances data from the external and internal thermostats and calculates the best time for boiler and associated plant to commence. In very large systems, typically office blocks, the daily adjusted plant running time is responsible for considerable fuel economies.

Energy management systems

All devices which contribute to energy saving and control, may be classified as part of an energy management system. The simplest time clock constitutes a system and at the other extreme, some very sophisticated computerised monitors may be deployed to contemporary intelligent buildings. Existing buildings rarely justify the expenditure, unless completely refurbished, and an energy survey/audit shown in Figure 3.28 will be essential to establish the viability.

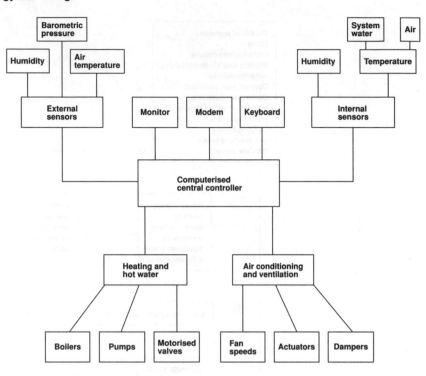

Figure 3.29 Energy management system

Figure 3.29 shows possible energy management system components and the network of system and environmental sensors. The information interface includes data relating to time of day, day of the week, building occupancy, meteorological information, internal environmental conditions, system state and efficiency of plant, etc. These details are processed to regulate the plant and equipment and to provide data for maintenance.

Pneumatic controls

Motorised valves are generally actuated by small electric motors. In many industrial situations where chemicals, paints, volatile fluids, plastics, etc. are regarded as a fire hazard, the building insurers and local fire officer may specify compressed air controls. Electrical compressor motors, sensors, etc. are housed outside the hazardous zone and motorised valves are fitted with pneumatic linear actuators shown in principle in Figure 3.30. These are either single or double action. Single acting requires air to extend the valve spindle, when the pressure is removed a spring retracts it. Double acting is more responsive, with extension and retraction responding to separate air supplies. In Figure 3.30, P1 supplies and P2 exhausts and vice versa.

Combined heat and power (CHP)

Combining heat and electricity generation into one production unit is a relatively new venture for modest size buildings and estates. Typical applications

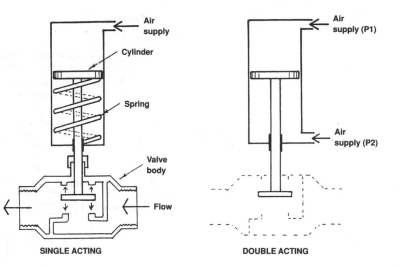

Figure 3.30 Pneumatic
linear actuators

include hospitals, hotels, country clubs, factories, etc. The concept has been applied to many remote buildings such as prisons for some time and large factories have also been independent of national power distribution. More recently, the initiative for power independence in the UK has been in response to the privatisation of electricity generation and distribution. Many corporations and sole building owners have installed their own gas- or oil-fired electricity generators using the by-product of surplus heat for hot water storage and heating. It is not yet viable for smaller dwellings, unless the system is shared in a district or between a community.

Power stations are only about 50 per cent efficient, with enormous energy wastage in the generator cooling water. This can be seen condensing above the massive cooling towers in national power plants. In CHP systems the hot water can be contained and reprocessed for distribution in mains to calorifiers and radiators, as shown in the schematic layout of Figure 3.31.

District, group and block heating

District heating is the generic term for a hot water supply system to several buildings or units (e.g. flats) within a large building. It may be part of a CHP system or just very large-scale independent hot water production and distribution.

Block systems apply to the central supply of hot water and heating from one plant source within a building and metered delivery to each self-contained unit. This eliminates the need for numerous boilers scattered throughout the building and associated problems of fuel supply and flue provision. Applications include blocks of flats, shopping complexes and office blocks.

Group systems have a centralised boiler plant serving several outbuildings. This is typical of hospitals, barracks, holiday camps, prisons, housing estates,

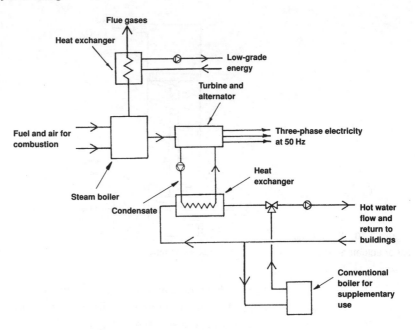

Flue gases

Heat exchanger

Low-grade energy

Turbine and alternator

Fuel and air for combustion

Three-phase electricity at 50 Hz

Heat exchanger

Steam boiler

Condensate

Hot water flow and return to buildings

Conventional boiler for supplementary use

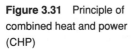

Figure 3.31 Principle of combined heat and power (CHP)

universities and other campus situations. Pipework is preferable underground in well-insulated ducts for easy maintenance, but in factories pipes are often suspended from the building's exterior. Depending on ownership, metered supply may be appropriate to each unit supplied.

District heating is an enlargement of group heating, applied to a municipality or town. Hot water for mains distribution is likely to come from a combination of sources. This could include CHP plant, conventional boilers, refuse burning and industrial heat-recovery units. Production and distribution of hot water will be under pressure (see Chapter 2) to ensure efficient delivery to strategically located heat exchangers. These process the pressurised water temperature from about 120 °C to a safer 80 °C at atmospheric pressure for domestic use. Factories and offices may use the higher temperature water for more efficient heat emission in convectors and air conditioners. Figures 3.32 and 3.33 show the principle of a two-pipe distribution, but the three-pipe system in Figure 3.34 is more effective for UK installations. The separate hot water and heating supplies satisfy the different seasonal demands. A common return is acceptable, although they too could be separated to create a four-pipe system.

Ducted warm air

The concept is similar to a fan-assisted convector (Figure 3.8), with ducts extending the warm air delivery to specific outlets. The central plenum may have either direct or indirect means of energy transfer. Direct systems use the

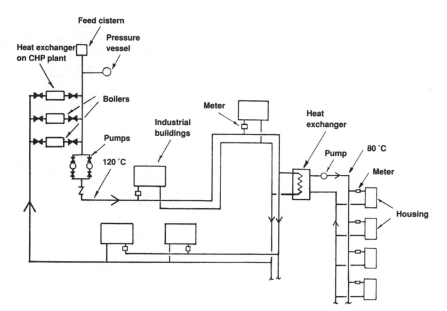

Figure 3.32 District heating, two-pipe distribution

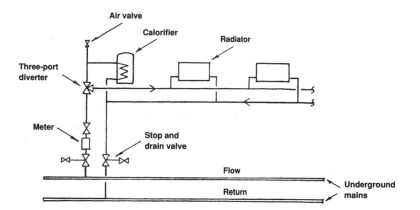

Figure 3.33 District heating, two-pipe supply

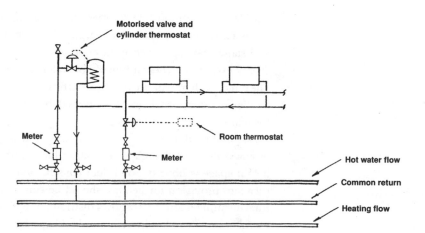

Figure 3.34 District heating, three-pipe supply

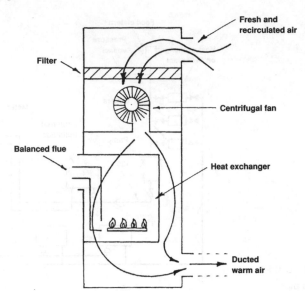

Figure 3.35 Warm air
heating unit – direct

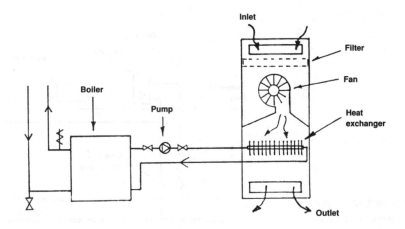

Figure 3.36 Warm air
heating – indirect

outside of the combustion compartment of a gas- or oil-fired boiler to heat a
fan-assisted airflow, before discharging through a system of ducting. Figure
3.35 shows typical direct application and Figure 3.36 illustrates an indirect
energy exchange, where a finned hot water element is located in the plenum.
The direct system will provide more efficient heat transfer, but because of flue
location on an external wall, the unit will be decentralised and less convenient
for uniform duct distribution.

   Distribution ducting is space-consuming and best located under suspended
or raised flooring. A high aspect ratio (duct width to depth proportions) will
help if space is limited. Solid floor construction is unlikely to be compatible,
therefore provision for heating should be considered in advance of architec-
tural detailing. Figure 3.37 shows distribution with galvanised sheet steel ducting, although flexible wire-reinforced tubing is a popular alternative, often used

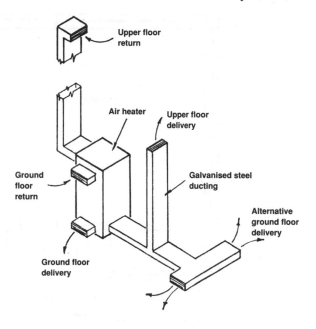

Figure 3.37 Warm air heating, distribution ducting

for main duct to terminal connections. Diffuser grills/terminals can be flush with the floor, wall or ceiling and have adjustable dampers to regulate airflow. Before distribution, recirculated air must mix with fresh air in proportions of no more than 3 : 1, respectively. Otherwise, oxygen depletion and stale air contamination will create an unhealthy environment.

Heating design

The objective is to match the thermal emission from radiators, to the thermal losses and energy transfer through the fabric of the building. This can be calculated by establishing a reasonable design base, i.e. internal temperatures to suit most occupants of 18–22 °C, depending on room function, and an external temperature of about –1 °C, depending on geographical location. In the UK, the external design criteria could range from –4 °C in the northerly extremes to 0 °C in the south-west. Although temperatures will fall below these on occasions, it is not considered economic to design radiators for extremes.

Figures 3.38 and 3.39 show the floor plans for a house. Given the following data, the radiator specification can be obtained:

Room height	2.4 m
Window height	1.2 m
Door height	2.0 m
Uniform internal design temperature	20 °C
External design temperature	–1 °C
Ventilation allowance	2 air changes/hour

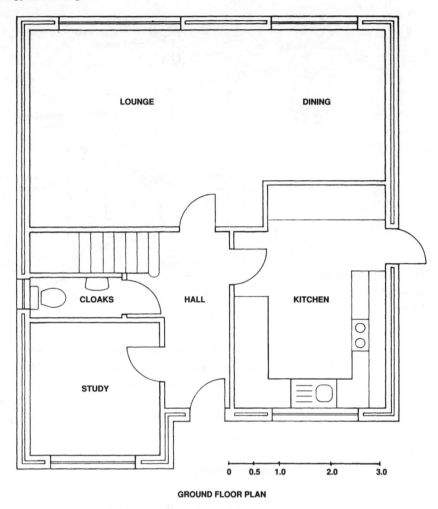

Figure 3.38 Detached house, ground floor plan

GROUND FLOOR PLAN

'U' values (W/m² K):

External wall	0.45
Floor	0.45
Ceiling/roof	0.25
Window	3.00
Door	2.5

The procedure is in three parts:

1. Calculate the heat lost due to air infiltration (A).
2. Calculate the heat lost through the fabric or structure of the building (B).
3. Sum (A) + (B) to establish the total heat loss (C).

Having found (C), reference to a radiator manufacturer's catalogue will indicate suitable emitters to complement this figure.

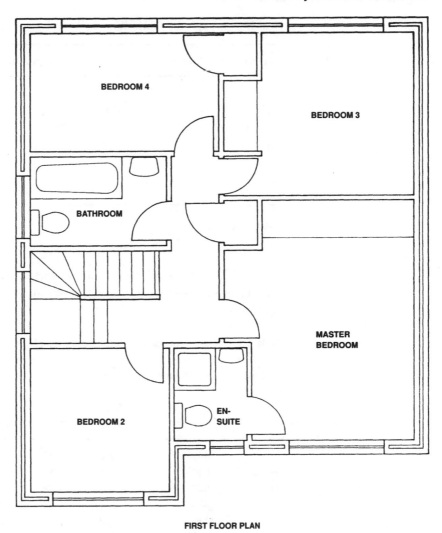

FIRST FLOOR PLAN

Figure 3.39 Detached house, upper floor plan

As an example, let us find the total heat loss for the study in Figure 3.38. Firstly, heat loss by ventilation (A). The following formula may be applied:

$$\text{Watts} = \frac{\text{Room volume} \times \text{ventilation allowance} \times \text{temperature difference}}{3}$$

Note: Temperature difference refers to the internal and external design differential.

$$\text{Watts} = \frac{(2.55 \times 2.65 \times 2.4) \times 2 \times (20 - -1)}{3}$$

$$= 227$$

Heat loss through the fabric (B):

Element	×	Area (m²)	×	'U'	×	Temperature difference	=	Watts
External wall		12.02		0.45		21		113.59
Window		2.02		3.00		21		127.26
Floor		6.76		0.45		21		63.88

$$\text{Total} = 304.73$$
$$\text{i.e. 305 W}$$

Total heat loss (C):

$$227 + 305 = \underline{532\ W}$$

In Table 3.3, sizes of available radiators extracted from a catalogue are given.

Using the same method, calculate the radiator sizes for some of the remaining rooms.

Table 3.3 Radiator catalogue extract. Single panel specification

Sections	Length (mm)	Height (mm)				
		300 Watts	400 Watts	500 Watts	600 Watts	750 Watts
12	480	303	388	470	550*	660
16	640	401	514	622	728	882
20	800	499	639	774	905	1096
24	960	596	764	925	1082	1310
28	1120	693	888	1075	1257	1523
32	1280	790	1011	1225	1432	1735
36	1440	886	1134	1374	1607	1946
40	1600	982	1257	1523	1781	2157
44	1760	1078	1380	1671	1954	2367
48	1920	1173	1502	1819	2128	2577
52	2080	1269	1624	1967	2301	2786

Note: Number of sections refers to the number of corrugations in the radiator. Final selection will depend on the available space, but an asterisk indicates the most likely.

Total heating design   To fully design a heating system, it is necessary to calculate the following:

- radiator sizes
- boiler rating
- pipe sizes
- pump specification.

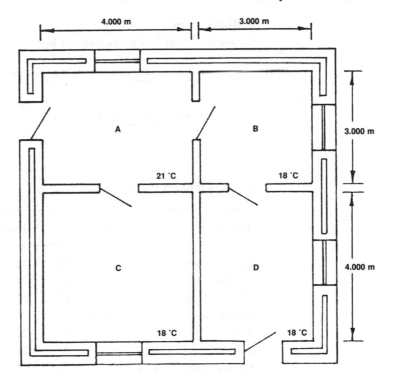

Figure 3.40 Heating
design, floor plan

This can be a long and involved procedure for most buildings, much simplified with computer-aided design packages. For illustration and explanation,
the simple structure shown in Figure 3.40 will demonstrate the concept and
numerical application.

Data:

Room height	2.5 m
Window area	1.5 m^2
Door area	2.0 m^2
External design temperature	–1 °C
Ventilation allowance	2 air changes/hour

'U' values (W/m^2 K):

External walls	0.6
Floor	0.3
Ceiling/roof	0.3
Internal walls	2.0
Doors	2.5
Windows	2.7

Radiator sizes

For room A, heat loss by ventilation is

$$\text{Watts} = \frac{\text{Room volume} \times \text{ventilation allowance} \times \text{temperature difference}}{3}$$

$$= \frac{(4 \times 3 \times 2.5) \times 2 \times (21 - -1)}{3}$$

$$= 440$$

Heat loss through the fabric:

Element	Area (m²)	'U'	Temperature difference	Watts
External wall	14	0.6	22	184.8
Window	1.5	2.7	22	89.1
External door	2.0	2.5	22	110.0
Internal door	2 No. × 2.0	2.5	3	30.0
Floor	12	0.3	22	79.2
Ceiling/roof	12	0.3	22	79.2
Internal walls	13.5	2.0	3	81.0
			Total =	653.3

Total heat loss:

$$440 + 654 \ (653.3 \text{ rounded up}) = 1094 \text{ W}$$

Similarly

$$\text{Room B} = 571 \text{ W}$$

$$\text{Room C} = 914 \text{ W}$$

$$\text{Room D} = 849 \text{ W}$$

Radiator specification from Table 3.3, using a consistent height of 400 mm:

Location	Design output (W)	Length (mm)	No. of sections	Output (W)
Room A	1094	1440	36	1134
B	571	800	20	639
C	914	1280	32	1011
D	849	1120	28	888
			Total =	3672 W

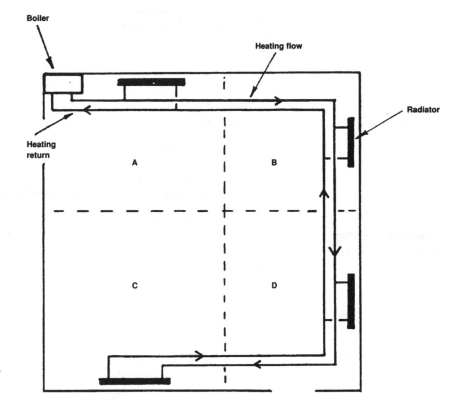

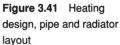

Figure 3.41 Heating design, pipe and radiator layout

Boiler rating

Allowing for a 75 per cent efficient boiler, the boiler output for heating purposes will be

$$3672 \times \frac{100}{75} = 4896 \text{ W}$$

It is usual for the boiler to provide hot water for storage as well as heating, therefore a separate calculation will be necessary (see Chapter 2). The full boiler specification will be the summation of both.

Pipe sizes

Pipe sizing depends very much on system layout. Figure 3.41 shows a possibility for this situation, with Figure 3.42 a schematic representation. For each pair of pipes, i.e. 1–4, it is necessary to calculate the mass flow rate in kg/s. Formula:

$$\text{kg/s} = \frac{\text{kW}}{\text{s.h.c.} \times \text{temperature difference (flow – return)}}$$

where s.h.c. is the specific heat capacity of water, normally taken at about 4.2 kJ/kg K.

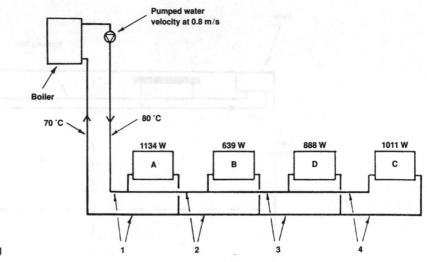

Figure 3.42 Heating design, schematic drawing

For pipes 1:

$$kg/s = \frac{1.134 + 0.639 + 0.888 + 1.011}{4.2 \times (80 - 70)} = 0.087$$

For pipes 2:

$$kg/s = \frac{0.639 + 0.888 + 1.011}{42} = 0.060$$

For pipes 3:

$$kg/s = \frac{0.888 + 1.011}{42} = 0.045$$

For pipes 4:

$$kg/s = \frac{1.011}{42} = 0.024$$

By referring to Figure 3.43, it is possible to coordinate the water velocity (min. 0.75 m/s, max. 1.0 m/s for copper tube) with the calculated mass flow rates. Thus,

> Pipes 1 are 15 mm outside diameter
> Pipes 2 are 12 mm outside diameter
> Pipes 3 are 10 mm outside diameter
> Pipes 4 are 8 mm outside diameter

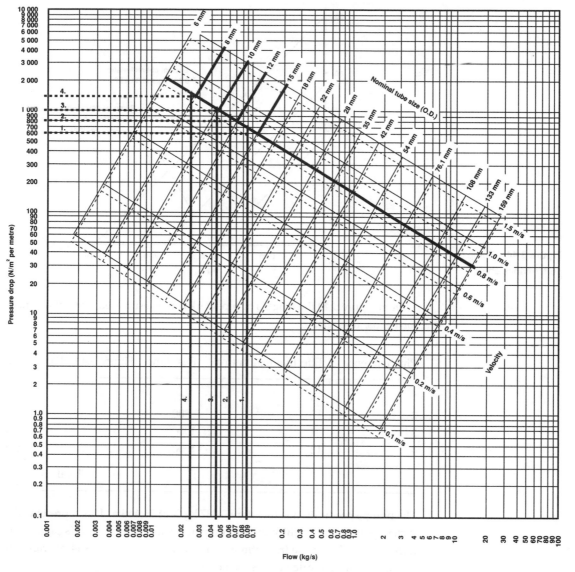

Figure 3.43 Heating design, pipe sizing and pump rating:
——————— 65 °C
- - - - - - - - - 115 °C

Pump specification

Figure 3.43 can also be used to determine the output of the circulating pump. Horizontal lines are drawn from the pipe diameter coordinates to the left of the page. This provides the resistance that the pumped water must overcome in units of N/m² or pascals per 1 m length of pipe.

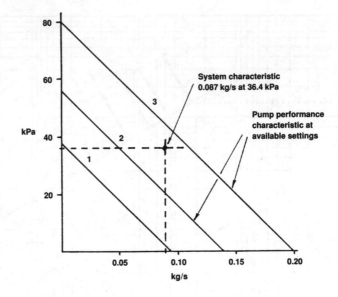

Figure 3.44 Pump performance chart

From Figure 3.40 the lengths of pipework can be estimated, with allowances for bends, tees, etc.

Pipe reference	Effective length (m)	Resistance (Figure 3.43) (Pa)	kPa
1	8	600	4.8
2	12	800	9.6
3	4	1000	4.0
4	12	1500	18.0
		Total =	36.4

The pump specification is therefore 0.087 kg/s at 36.4 kPa. This represents the system performance, which can be compared with manufacturer's pump performances to determine the optimum pump selection. Figure 3.44 provides an example using the calculated specification.

4 Ventilation

Ventilation is simply defined as the process of changing air in an enclosed space. A proportion of air within the enclosed space should be continuously withdrawn and replaced by fresh air. This must be drawn in from a clean external source, generally at as high an elevation as practical, particularly where more polluted air occurs at the relatively low levels found in congested towns and cities.

Ventilation is needed to maintain air purity, i.e.:

- preservation of oxygen content – this should be maintained at approximately 21 per cent of air volume
- removal of carbon dioxide
- control of humidity – between 30 and 70 per cent relative humidity (RH) is acceptable for human comfort
- prevention of heat concentrations from machinery, lighting and people
- prevention of condensation
- dispersal of concentrations of bacteria
- dilution and disposal of contaminants such as smoke, dust, gases and body odours
- provision of freshness – an optimum air velocity lies between 0.15 and 0.5 m/s.

Ventilation requirements

Control of ventilation rates is influenced by various authorities and codes of practice, some statutory and others recommendations. The latter in the form of British Standard (BS) design guidelines are often used as the criteria for establishing mandatory applications.

The Factories Act and the Offices, Shops and Railway Premises Act require 'adequate' standards of ventilation. This gives the local authority health and safety executive flexibility to interpret each situation accordingly. BS 5720 recommendations for mechanical ventilation are a likely resource for determining acceptable criteria. A variation is produced in Table 4.1, with a calculation to convert cubic metres of fresh air per hour per person to the more practical ventilation allowance or number of air changes per hour. Table 4.2 provides approximate air changes for various situations.

Table 4.1 Fresh air supply rates

Type of space	Recommended m³/hour per person
Factory	18–30
Open-plan office	
Shops	
Department store	
Supermarket	
Theatre	
Cafeteria	30–43
Dance hall	
Hotel bedroom	
Laboratories	
Private offices	
Residential	
Cocktail bar	43–65
Function room	
Luxury residential	
Restaurant/commercial dining room	
Boardroom	65–90
Executive office	
Conference room	
	Recommended m³/hour per m² of floor area
Corridors	5
Domestic kitchen	36 (see also Table 4.3)
Commercial kitchen	72
Sanitary accommodation	36 (see also Table 4.3)

Note: To convert to air changes per hour, divide selected figures by room volume, and multiply by the number of occupants. For example a private office of 30 m³ volume designed for two people:

$$\frac{43}{30} \times 2 = 2.86 \text{ air changes per hour}$$

The Building Regulations, Approved Document F, defines the minimum standards acceptable in new housing to achieve an objective of one-half to one air change per hour. This is sufficient to complement high insulation standards while preventing condensation. Table 4.3 establishes the criteria and Figure 4.1 shows the application.

Table 4.2 Approximate air change rates

Accommodation	Air changes per hour
Offices – above ground	2–6
Offices – below ground	10–20
Factories – large, open	1–4
Factories/industrial units	6–8
Workshops with unhealthy fumes	20–30
Fabric manufacturing/processing	10–20
Kitchens – above ground	20–40
Kitchens – below ground	40–60
Public lavatories	6–12
Boiler accommodation/plant rooms	10–15
Foundries	8–15
Laboratories	10–12
Hospital operating theatres	<20
Hospital treatment rooms	<10
Restaurants	10–15
Smoking rooms	10–15
Storage/warehousing	1–2
Assembly halls	3–6
Classrooms	3–4
Domestic habitable rooms	Approx. 1
Lobbies/corridors	3–4
Libraries	2–4

Figure 4.1 Ventilation of dwellings:
(A) window with minimum openable area = 1/20 floor area;
(B) mechanical ventilation of at least 6 l/s;
(C) mechanical ventilation of at least 15 l/s, or PSV;
(D) mechanical ventilation of at least 60 l/s (30 l/s if in cooker hood) or PSV;
(E) trickle ventilation of at least 4000 mm²;
(F) trickle ventilation of at least 8000 mm²

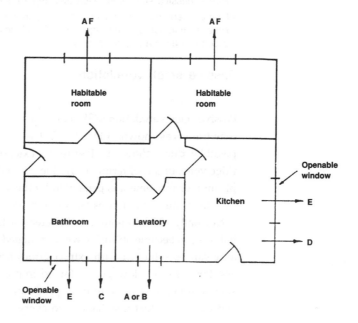

Table 4.3 Building Regulation requirements for domestic ventilation

Accommodation	Ventilation provision		Extract rate (l/s)
	Rapid	Background (mm²)*	
Bathroom (with or without WC)	Openable window	4000	15 or PSV[†]
Habitable room	Openable window or ventilation equivalent to 1/20 floor area	8000	—
Kitchen	Openable window	4000	30 in cooker hood, 60 elsewhere or PSV
Sanitary accommodation (if separate from bathroom)	Openable window or ventilation equivalent to 1/20 floor area or mechanical extract of 6 l/s or PSV	4000	—
Utility room	Openable window	4000	30 or PSV
Common space	Ventilation opening of 1/50 floor area or extract fan providing at least one air change per hour		

* Background ventilation may be by trickle vents set in window frames.

[†] PSV = passive stack ventilation (see Figure 4.2).

PSV only appropriate if an internal room, i.e. no external wall.
  Note: 15 min. fan overrun required for fans in internal bathrooms, kitchens, sanitary accommodation and utility rooms.

Passive stack ventilation

Passive stack ventilation (PSV) combines with trickle ventilation through window frames as shown in Figure 4.2. The simple vertical (or as near vertical as possible) ducts shown in Figure 4.3 extend from kitchen and bathroom to ridge vents or a roof terminal. Trickle ventilation provides the natural air supply and air movement is generated by the stack effect (see Figure 4.6), created by temperature differences between inside and outside of the building.

Normally 100 mm nominal diameter plastic or flexible ducting is adequate, but this can become obtrusive when insulated and boxed in. Enhanced performance can be achieved with extract grilles activated by humidity sensors and assistance from a low power fan to improve airflow. Assisted installations, shown in Figure 4.4, are more appropriate in blocks of flats to eliminate long and obtrusive duct installations to every compartment.

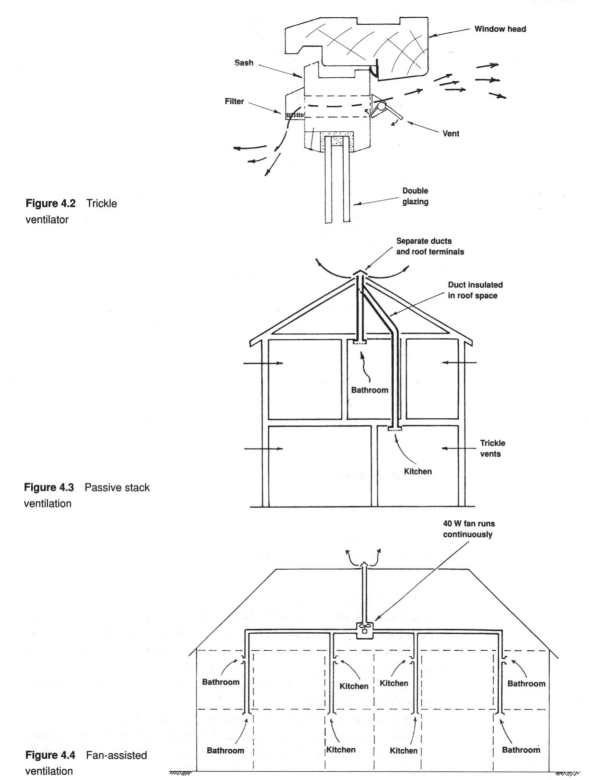

Figure 4.2 Trickle ventilator

Figure 4.3 Passive stack ventilation

Figure 4.4 Fan-assisted ventilation

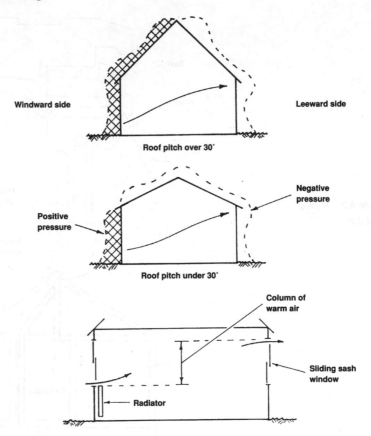

Figure 4.5 Wind pressure and natural ventilation

Natural ventilation

Natural ventilation relies on either:

- wind direction and pressure, or
- the stack effect of warm air rising within a building, while cooler air exists outside.

The former is dependent on a consistent south-westerly wind direction, with distribution shown in Figure 4.5. Effectiveness is variable and limited by the fickle nature of the weather. A number of Victorian and Edwardian school buildings, hospitals and community rooms can still be found built to these principles. Long and tall in proportions, to take full advantage of the cross-flow of air and the associated column of warmth derived from a radiator to the windward elevation.

Other sources of air infiltration included gaps around doors and windows, chimneys, gaps between floorboards and skirtings. Contemporary buildings are well sealed and insulated, promoting a high condensation risk, hence the need for purpose-made ventilation as detailed in Table 4.3.

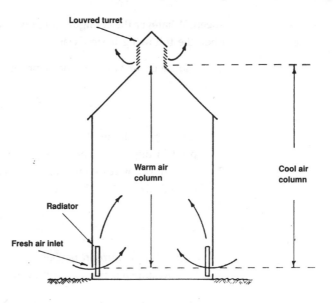

Figure 4.6 Stack effect
ventilation

Stack-effect heating and ventilation is another well-established means of air movement in tall narrow buildings, typical of mills and other historic industrial buildings. Figure 4.6 shows the principle, with warm air rising between purpose-made voids in the different floor levels. By contemporary standards it would be unacceptable on grounds of fire spread potential and energy wastage, but it has been effective in modern shopping malls and atrium enclosed areas. The surplus warmth generated at low levels from people, lighting and shops follows the same stack-effect extraction principles in many modern commercial developments. Efficiency depends on at least a 10 °C higher temperature differential between inside and out, hence the limitations of stack-effect ventilation in the summer months, when ironically it is most needed.

Mechanical ventilation

Since electricity became readily available in the early part of the twentieth century, fan-assisted movement of air has largely superseded the unreliable natural systems. Components include:

- fan
- filters
- ductwork
- fire dampers
- diffusers.

Fans Fans provide the motive power for air movement by imparting static energy or pressure and kinetic energy or velocity. A fan's capacity for air movement depends on its characteristics such as type, size, shape, number of blades and speed. Whatever the configuration, given constant air density, fan capabilities obey the following basic laws:

1. Volume of air varies in direct proportion to the fan speed; i.e.

$$\frac{Q_2}{Q_1} = \frac{N_2}{N_1}$$

where Q = volume of air in cubic metres per second (m^3/s) and N = fan speed in impeller revolutions per minute (rpm).

2. Pressure of, or resistance to, air movement is proportional to fan speed squared, i.e.

$$\frac{P_2}{P_1} = \frac{(N_2)^2}{(N_1)^2}$$

where P = pressure in pascals.

3. Air and impeller power is proportional to fan speed cubed, i.e.

$$\frac{W_2}{W_1} = \frac{(N_2)^3}{(N_1)^3}$$

where W = power in watts or kilowatts.

For example, in a ducted air movement system, a fan of 2 kW power discharges 4 m^3/s with impellers rotating at 1000 rpm to produce a pressure of 250 Pa. If the fan/impeller speed increases to 1250 rpm, calculate the revised performance data.

1. $\dfrac{Q_2}{Q_1} = \dfrac{N_2}{N_1}$

$$\frac{Q_2}{4} = \frac{1250}{1000} \quad \text{therefore } Q_2 = 5\,m^3/s.$$

2. $\dfrac{P_2}{P_1} = \dfrac{(N_2)^2}{(N_1)^2}$

$$\frac{P_2}{250} = \frac{(1250)^2}{(1000)^2} \quad \text{therefore } P_2 = 390\,Pa.$$

3. $\dfrac{W_2}{W_1} = \dfrac{(N_2)^3}{(N_1)^3}$

$$\frac{W_2}{2} = \frac{(1250)^3}{(1000)^3} \quad \text{therefore } W_2 = 3.9\,kW.$$

It should be noted that fans are not totally efficient and the following formula may be applied to determine the percentage:

$$\text{Efficiency} = \frac{\text{Total fan pressure} \times \text{air volume}}{\text{Absorbed power (watts)}} \times \frac{100}{1}$$

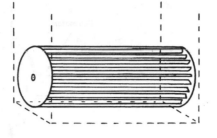

Figure 4.7 Tangential or
cross-flow fan

So, for the previous example,

$$\text{Efficiency} = \frac{390 \times 5}{3900} \times \frac{100}{1} = 50 \text{ per cent}$$

Types of fan

There are four categories of fan suitable for air movement in ventilation systems:

1. Cross-flow or tangential
2. Propeller
3. Axial flow
4. Centrifugal.

Cross-flow or tangential

This is a long cylindrical unit with peripheral impellers shown in Figure 4.7. Efficiency is limited to about 45 per cent, restricting its application to portable units and fan coil convectors of the type shown in Figure 3.9.

Propeller

A propeller fan comprises several steel or plastic blades mounted at right angles to a central boss. In free-standing form they can be used on office desks, but are most frequently applied domestically to voids in kitchen and bathroom walls. They do have a high volume capacity, which is ideal for extract applications to public lavatories, small canteens, workshops, etc. Low-pressure potential with efficiency less than 40 per cent makes it unsuited to ducted systems. Figure 4.8 shows an industrial version, complete with wind protection guard to prevent back flow of stale air.

Axial flow

These consist of several aerofoil cross-section blades mounted on a motor-driven central shaft. The whole unit is located in a circular housing for adaptation to ductwork. High efficiencies of up to 75 per cent are achieved by closeness of fit between blades and housing and the aerodynamic design, twist and pitch angle of the blades. Figure 4.9 shows how compact the unit is, with integral variable-speed motor. Larger versions have belt-driven impellers

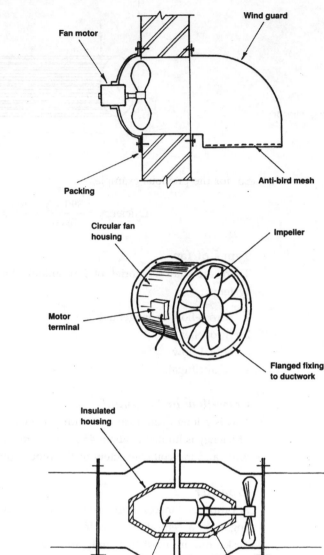

Figure 4.8 Wall-mounted propeller fan

Figure 4.9 Axial-flow fan

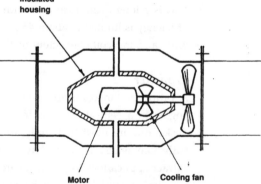

Figure 4.10 Bifurcated axial flow fan

from an external motor. The bifurcated variation in Figure 4.10 has an insulated housing, to protect the fan-cooled motor in greasy, hot and corrosive gas situations.

Centrifugal

These fans house an impeller rotating in an involute or scroll-shaped casing. Air is drawn in at right angles before discharging radially under centrifugal force through the delivery ductwork. Small fans have an integral motor mounted centrally within the impeller, but larger models associated with high

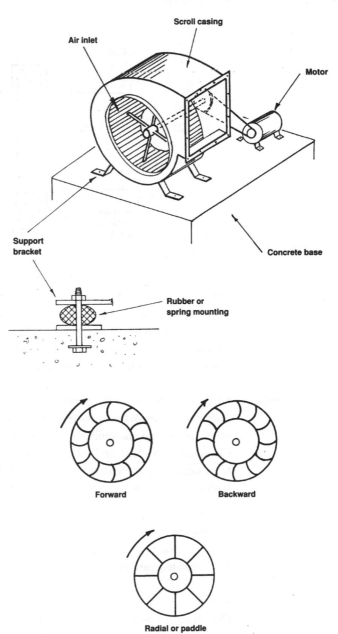

Figure 4.11 Centrifugal fan

Figure 4.12 Centrifugal fan impellers

pressure and long deliveries have an external motor and pulley block system of gearing. Figure 4.11 shows this arrangement, which has considerable flexibility in use, i.e. a variable-speed motor, gearing through the pulley blocks and interchangeable impellers seen in Figure 4.12.

The forward curved blade is best suited to low and medium constant pressures, as a high system resistance could overload the fan with this impeller. Backward curves apply to high and variable resistances, such as that found in

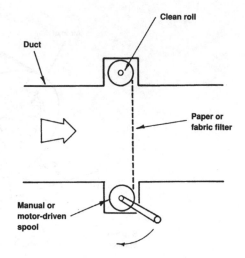

Figure 4.13 Roll filter

high-velocity ventilation systems. Paddle or flat blades are appropriate where dusty, dirty or other suspended matter is present in the air. More plant room space is needed for these fans than the axial flow fans, but their inherent adaptability, quietness of operation and efficiency rate of up to 75 per cent make them more favourable for larger installations.

Filters Filters are applied at the point of air entry into a ventilation system. Their purpose is to remove suspended particles, contaminants and odours that would otherwise offend the building's occupants and contribute to deterioration of the ventilation plant and internal finishes. There are various types, ranging from simple paper elements to sophisticated electronic devices, corresponding with cost and degree of efficiency. Classification is in four categories:

1. Dry
2. Viscous
3. Electrostatic
4. Activated carbon.

Dry

Dry filters can be produced from paper, fine woven fabrics, foamed plastics or glass fibres. Both paper and fabric filters are produced in rolls for suspension in the air draught. Figure 4.13 shows the arrangement with choice of automatic or manual rotation at necessary intervals. Automatic rotation can be effected when air resistance operates a pressure-sensitive switch to engage the motor-driven spool. Other types of dry filter are available as disposable elements. These fit into a purpose-made compartment at the air intake. Figure 4.14 shows variations, some with a V-formation to increase the surface area. These are usually replaced at set intervals depending on air quality, but may be vacuumed two or three times before rejection.

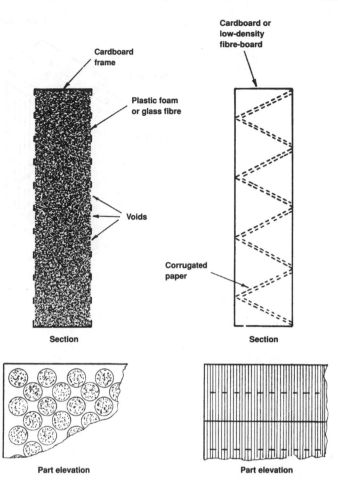

Figure 4.14 Disposable element filters

Viscous

Viscous or wet filters comprise a bank or rows of corrugated metal sheets with surface coated in a non-flammable, non-toxic odourless oil. Figure 4.15 shows the arrangement and an alternative perforated element type of filter. These are most suited to industrial situations, where heavy air contamination is apparent and the suspended particles will adhere to the oily surface. The plates or element are cleaned in dilute caustic soda or an oil-diluting medium for replacement at specified intervals.

Electrostatic

Electrostatic filters or precipitators are a very expensive and extremely efficient means of removing fine particles, pollens and smoke from the air. They are normally installed with a pre-filter to remove larger dust particles before the main unit. This contains power pack, ioniser and rows of oppositely charged plates as shown in Figure 4.16. The ioniser positively charges suspended particles before conveyance in the airstream through the metal plates. The

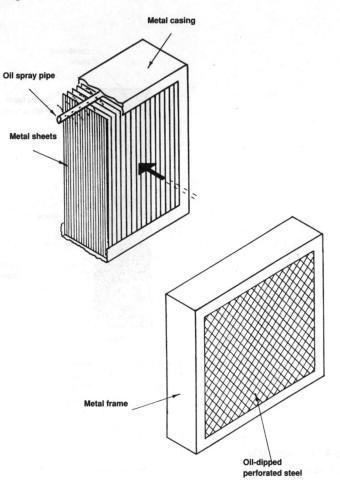

Metal casing

Oil spray pipe

Metal sheets

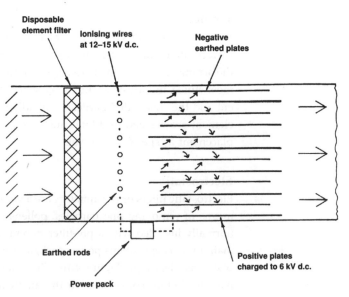

Metal frame

Oil-dipped
perforated steel

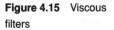

Figure 4.15 Viscous
filters

Disposable
element filter

Ionising wires
at 12–15 kV d.c.

Negative
earthed plates

Earthed rods

Power pack

Positive plates
charged to 6 kV d.c.

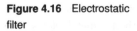

Figure 4.16 Electrostatic
filter

positively charged plates repel the dust to the earthed negative plates, which will require periodic cleaning.

Activated carbon

Activated carbon filters are very absorbent and specifically designed for use in greasy, odorous atmospheres such as that produced from commercial food-processing units. They are disposable elements with glass fibrous matting containing the coconut shell charcoal granules. Location is often within a cooker hood shown in Figure 4.17, to prevent grease penetrating and lining the extract ductwork. In these situations a bifurcated axial flow fan would be the most appropriate choice.

Ductwork Ductwork is produced in circular, square or rectangular cross-sections (shown in Fig. 4.18) in several different materials as listed in Table 4.4. Circular ductwork

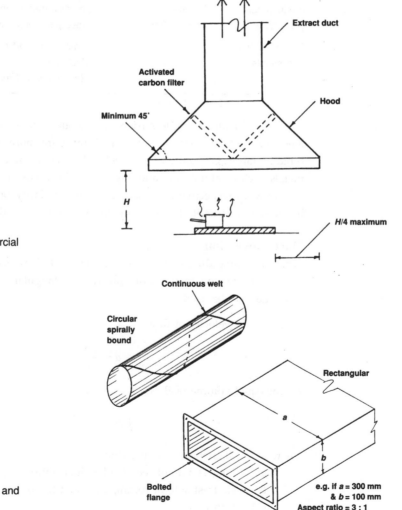

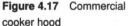
Figure 4.17 Commercial cooker hood

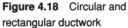
Figure 4.18 Circular and rectangular ductwork

Table 4.4 Ductwork data

Material	Min. thickness (mm)	Application
Galvanised steel	0.6	Low velocity (<10 m/s) Low pressure (<500 Pa)
	0.8	Velocity >10 m/s Pressure >500 Pa Spirally bound (Figure 4.18)
Aluminium	0.8	Low velocity
uPVC	3.0	Low velocity Temperature range −10 ˚C to +40 ˚C
Polypropylene	3.0	Low velocity Temperature range −10 ˚C to +40 ˚C
Resin-bonded glass fibre	Variable, approx. 3.0	Intrinsic thermal insulation and acoustic absorption properties
Flexible fabric and metals		Limited to short connections from main duct to terminal. Recommended max. length 6× diameter

is more efficient, having less frictional resistance to airflow, but for convenience, rectangular ducts of high aspect ratio are more easily fitted into the
building fabric. Galvanised mild steel is the most common material, but other
metals such as aluminium and copper have occasional use for external and
feature work respectively. uPVC is gaining popularity, but poor performance
in fire may restrict its use and suitability in some situations.

Duct conversion

When designing ductwork, most formulae and charts express the section as
circular. To convert to the more practical rectangular profile, the following
formulae may be used:

- for equal velocity of flow,

$$d = \frac{2ab}{a + b}$$

- for equal volume of flow,

$$d = 1.265 \times \left[\frac{(ab)^3}{a + b} \right]^{0.2}$$

where d = diameter of circular duct (mm)
  a = longest side of rectangular duct (mm)
  b = shortest side of rectangular duct (mm)
  0.2 = fifth root.

For example, a 450 mm diameter duct converted to rectangular profile of aspect ratio 2 : 1 ($a = 2b$). For equal velocity of flow:

$$d = \frac{2ab}{a + b}$$

$$450 = \frac{2 \times 2b \times b}{2b + b}$$

$$450 = \frac{4b^2}{3b}$$

$$450 = \frac{4b}{3}$$

$$b = \frac{3 \times 450}{4}$$

Therefore $b = 337.5$ mm and $a = 2b = 675$ mm.

For equal volume of flow:

$$d = 1.265 \times \left[\frac{(ab)^3}{a + b}\right]^{0.2}$$

$$450 = 1.265 \times \left[\frac{(2b \times b)^3}{2b + b}\right]^{0.2}$$

$$450 = 1.265 \times \left[\frac{(2b^2)^3}{3b}\right]^{0.2}$$

From this, $b = 292$ mm and $a = 2b = 584$ mm.

As the arithmetic is rather tedious, a much simpler but less accurate conversion can be obtained from the chart depicted in Figure 4.19.

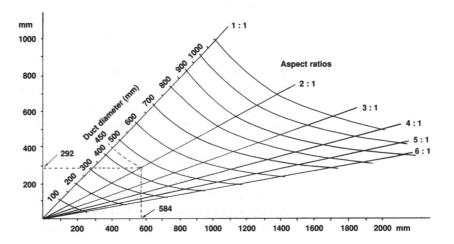

Figure 4.19 Circular to rectangular ductwork conversion chart

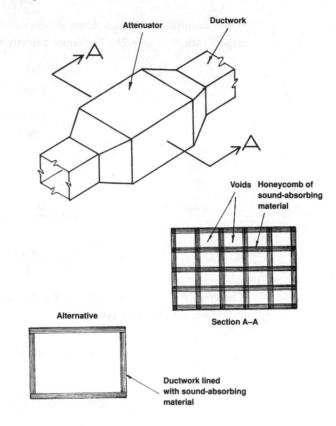

Figure 4.20 Sound attenuation

Noise control

Noise from air turbulence, fan and associated vibration must be controlled for the comfort of the building users. Secondary disturbance may also be apparent from sharp edges and bends in the ductwork, as well as noise around dampers. Sound transmittance will also depend on the perimeter and length of ductwork, as some reduction is achieved by the pressure waves being damped down on the duct walls. Long ducts are more effective at sound damping than those having a large area. If residual sound is likely to be a nuisance, the ductwork may be continuously lined with an absorbing material such as low-density fibre-board. Alternatively, purpose-made attenuators of the type shown in principle in Figure 4.20, can be provided at strategic intervals. In all systems, a flexible connection of reinforced PVC should be provided next to the fan housing. Minimal distraction from noise is possible by adhering to the velocity and pressure-drop guidelines in Table 4.5.

Volume and direction control

Splitters and dampers are used in ductwork to provide air direction and regulation. Splitters are blades or vanes fixed in the airstream to direct medium to high-velocity air around bends and offsets. They control flow conditions to reduce turbulence and associated noise. Dampers are rotating blades used to

Table 4.5 Recommended maximum ducted air velocities and resistances for acceptable levels of noise

Application	Maximum air velocity (m/s)	Maximum resistance or pressure drop (Pa/m)
Extremely quiet situations such as reading rooms, recording studios and operating theatres	2.5	0.4
Fairly quiet locations, e.g. church, dwellings, private rooms, offices, hospital wards, commercial premises, theatres, restaurants, public buildings, classrooms and conference facilities	6.0	0.6
Less critical situations, such as exhibition centres, factories, workshops, gyms, departmental stores, cafes/fast food centres, warehousing, etc.	10.0	0.8

regulate the airflow through the ductwork. They can be manually or automatically actuated to suit variable volumes and velocities at outlets. They are also used to control mixing of different qualities and properties of air, as shown in Figure 4.21 where recirculated air and fresh air combine to enter the processing unit.

Fire dampers The Building Regulations require multi-occupancy (flats), commercial and industrial use buildings to be compartmented to contain the spread of fire. Wherever ductwork penetrates compartment walls, floors or ceilings it must incorporate provision for automatic closure in the event of fire. Various types of fire damper exist; some shown in Figure 4.22 will satisfy building legislation, but final selection and type approval are likely to made by the local fire officer and the building's insurers. Gravity and spring-balanced devices with a soldered fusible link set at 70 °C are widely used, but large volumes of smoke can still permeate through the ductwork before sufficient heat builds up to release the damper. Smoke-detection devices are preferable and these can be used to operate a solenoid or electromagnetic shutter activator.

Diffusers Diffusers range from simple perforated plates and grilles to the more complex and efficient coned air distributors shown in Figure 4.23. The design and selection must achieve the appropriate amount of air distribution and throw for the given situation. Some examples are illustrated in Figure 4.24. Primary air from the ventilation system will impart into the room at sufficient velocity

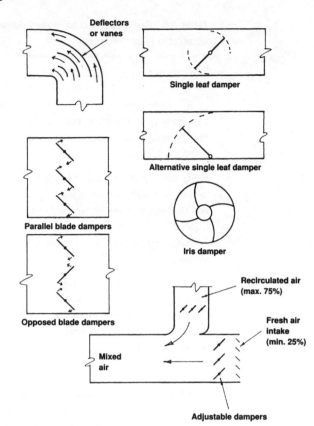

Figure 4.21 Air movement control

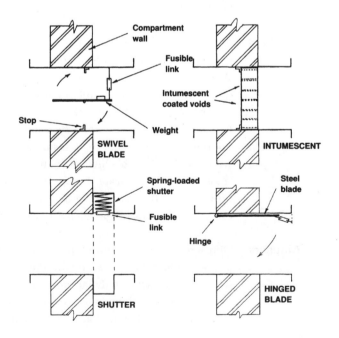

Figure 4.22 Fire dampers

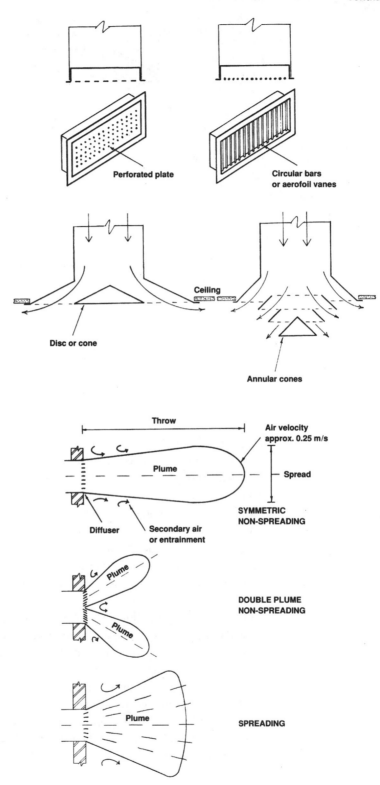

Figure 4.23 Grilles and diffusers

Figure 4.24 Diffuser airflow patterns

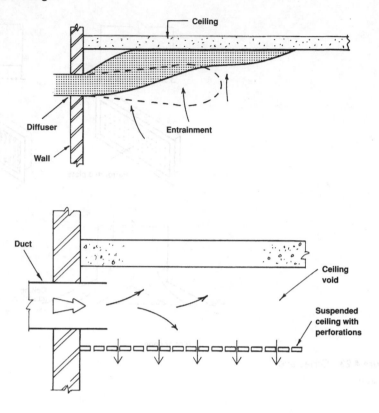

Figure 4.25 Coanda effect

Figure 4.26 Plenum ceiling

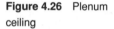

(not less than 0.75 m/s) to overcome the resistance of the surrounding air. As the airstream velocity reduces over distance and its temperature comes close to room temperature, it increasingly blends with the secondary air up to the throw distance. This corresponds to a reduced velocity of about 0.25 m/s. Movement of secondary air into the plume is called entrainment and this usefully reduces the possibility of stale air zones and hot spots, provided diffusers are strategically located. Correct selection of diffuser relative to discharge air velocity is also important to reduce unwanted draughts. These will be apparent if the throw is too great and the plume hits walls, beams and other obstructions. Care must also be observed where diffusers are located in adjacent walls and ceilings. The plume effect shown in Figure 4.25 is called a coanda and is created by restricted air and pressure at the adjacent surface due to limited access for air to replace the entrained air above the plume. Some designers like this effect as it extends the plume to run along a ceiling and given sufficient velocity, down the opposing wall as well.

As an installation economy and as a means of evenly distributing air, suspended ceilings and raised floors can be used as plenum chambers. The concept is shown in Figure 4.26, where a perforated false ceiling provides overall air permeation.

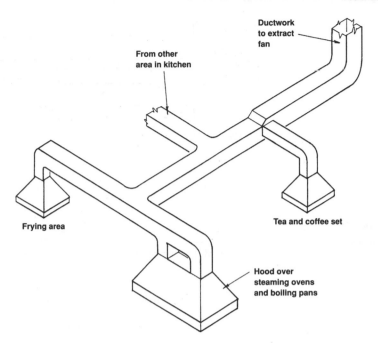

Figure 4.27 Extract ventilation to a commercial kitchen

Systems

Mechanical systems of ventilation are conveniently categorised:

1. Mechanical extract/natural supply
2. Mechanical supply/natural extract
3. Combined mechanical extract and supply.

Mechanical extract/ natural supply

The simplest application is provision of a propeller fan over a void in an external wall of a kitchen, bathroom or similarly unpleasant or contaminated air situation. More advanced systems have ductwork attached to the extract fan, with hoods or outlet grilles strategically located. The scheme shown in Figure 4.27 is typical of that required for a commercial kitchen, with replacement air from purpose-made vents or louvred grilles in the external wall. Figure 4.28 shows an application to assembly halls and lecture theatres. Ductwork is accommodated within a suspended ceiling, with fresh air inlets in peripheral locations. This arrangement encourages throughflow of clean air to displace the dusty, smoky stale atmosphere. The mechanical extract requirements for sanitary accommodation defined in Table 4.3 and Figure 4.1 are applied to blocks of flats by combining extract ducting with staggered or shunt ducts serving each compartment. Figure 4.29 indicates the principle, with the offset shunt preventing the cross-flow of noise, odours and smoke between adjacent compartments.

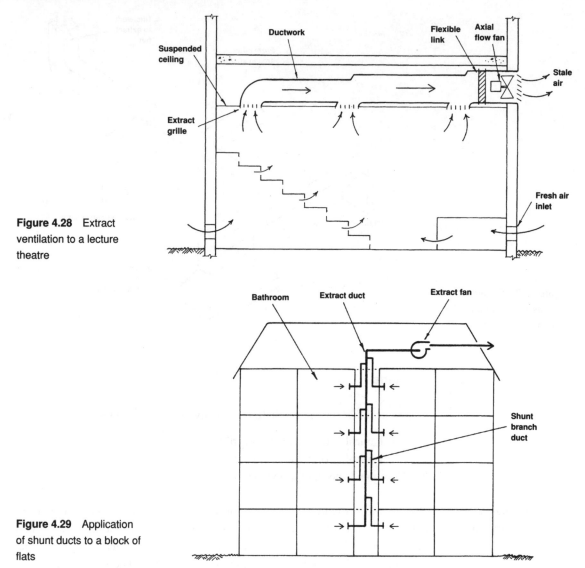

Figure 4.28 Extract ventilation to a lecture theatre

Figure 4.29 Application of shunt ducts to a block of flats

Mechanical supply/ natural extract

In essence this is the previous system with the fan reversed to deliver fresh air. Careful design and strategically located, restricted outlets will create slight internal pressurisation and direction for the stale air to permeate through the building. During winter the air will need preheating. Figure 4.30 shows an established plenum system with heat exchanger battery and filter positioned at the intake.

Combined mechanical extract and supply

By combining (1) and (2) above, the best possible means of ventilation is achieved. The extract fan should be smaller than the inlet, to encourage slight air pressurisation. Sealed windows and self-closing doors are needed to complement the efficiency of the system and to reduce draughts, dust and noise

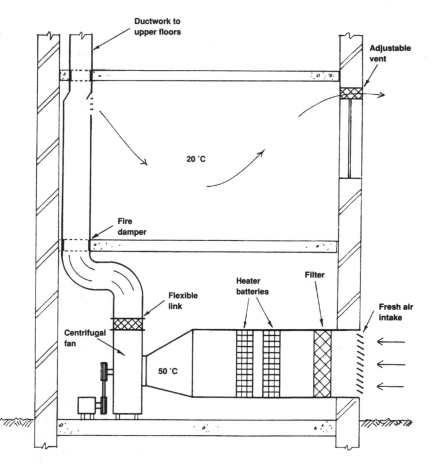

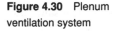

Figure 4.30 Plenum ventilation system

penetration. Figure 4.31 shows the concept, with the economies of recirculated air and lighting extract grilles.

Ventilation design

After determining the location of fan housing, ductwork and diffusers/grilles, the size of ducts can be calculated by a variety of means. Theoretically, the pressure drop or resistance to airflow in each section of ductwork should be much the same, to provide a perfectly balanced system. In practice this is difficult to achieve, as ductwork is rarely straight and uninterrupted, having to offset beams, columns and other structural priorities. Secondary balancing by dampers is invariably necessary to facilitate successful commissioning. When designing the ductwork, due regard should be given to the air velocity parameters and resistances recommended in Table 4.5. Procedure depends on the designer's priorities listed below. Either method is quite adequate for approximating ductwork sizes and fan rating:

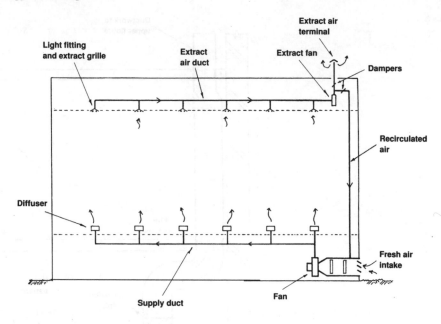

Figure 4.31 Combined mechanical extract and supply

- **Equal velocity** – the designer selects the same air velocity for use throughout the system.
- **Velocity reduction** – the designer selects variable velocities appropriate to each section or branch of ductwork, or
- **Equal friction** – the air velocity in the main duct is selected and the size and friction determined from a design chart (Figure 4.33). The same frictional resistance is used for all other sections of ductwork.

The simple extract ventilation system shown in Figure 4.32 will be used to illustrate the preceding methods of design, with the following formula for calculation of air volume flow rate:

$$Q = \frac{\text{Room volume} \times \text{air changes per hour}}{\text{Time in seconds}}$$

where Q is in cubic metres per second. From Figure 4.32, room volume = 480 m³; from Tables 4.1 and 4.2, air changes per hour = 6, therefore

$$Q = \frac{480 \times 6}{3600} = 0.8 \text{ m}^3/\text{s}$$

Equal velocity method Selected air velocity throughout the system (ducts A and B) is 5 m/s, from Table 4.5. Given that Q, the quantity of air = 0.8 m³/s is equally extracted through each grille, duct A will convey 0.8 m³/s and duct B 0.4 m³/s.

From the design chart in Figure 4.33 and extracted to Figure 4.34,

Duct A = 450 mm diameter

Duct B = 320 mm diameter

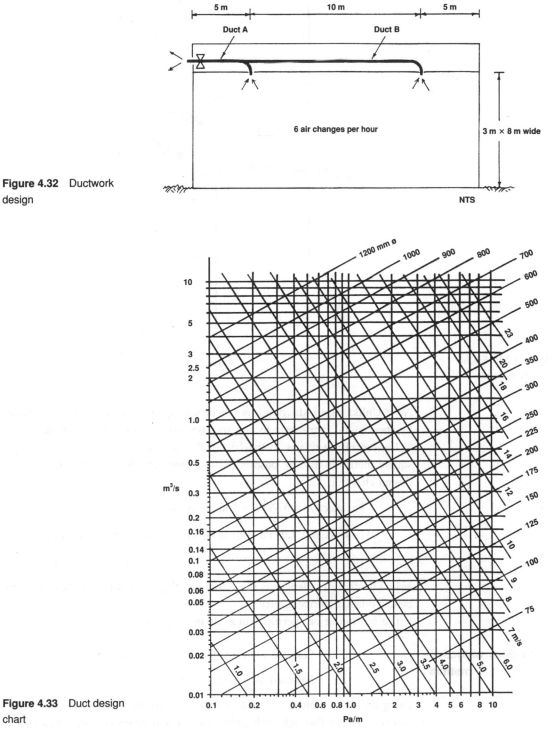

Figure 4.32 Ductwork design

Figure 4.33 Duct design chart

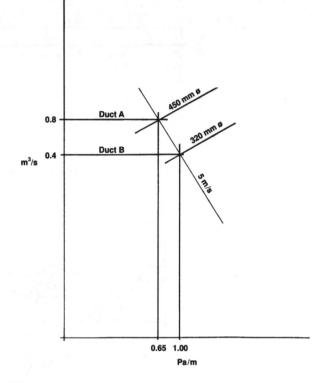

Figure 4.34 Equal
velocity method

Note: Unlike pipes, ductwork is not produced in stock sizes. It can be made by sheet metal fabricators to any specified diameter or cross-section.

The fan rating relates to the frictional resistance obtained in newtons per square metre (N/m^2) or pascals (Pa) per unit length of ductwork, from the design chart:

$$\text{For duct A} = 0.65\,\text{Pa} \times 5\,\text{m effective duct length} = 3.25\,\text{Pa}$$

$$\text{For duct B} = 1.00\,\text{Pa} \times 10\,\text{m effective duct length} = \underline{10.00\,\text{Pa}}$$

$$\text{Total} = 13.25\,\text{Pa}$$

Note: Effective duct length is the actual length plus additional allowances for bends, offsets, dampers, etc.

The fan rating or specification is $0.8\,\text{m}^3/\text{s}$ at 13.25 Pa.

Velocity reduction method

Selected air velocity in duct A = 6 m/s

Selected air velocity in duct B = 3 m/s

Using the air volume flow rate formula as before, and applying equal extraction of $0.4\,\text{m}^3/\text{s}$ through each grille, duct A will again convey $0.8\,\text{m}^3/\text{s}$ and duct B, $0.4\,\text{m}^3/\text{s}$. The design chart in Figure 4.35 shows that for the different velocities, ducts A and B are both coincidentally 420 mm diameter.

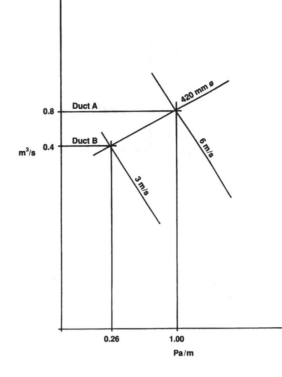

Figure 4.35 Velocity reduction method

$$\text{Friction in duct A} = 1.00\,\text{Pa} \times 5\,\text{m} = 5.0\,\text{Pa}$$

$$\text{Friction in duct B} = 0.26\,\text{Pa} \times 10\,\text{m} = 2.6\,\text{Pa}$$

$$\text{Total} = 7.6\,\text{Pa}$$

The fan rating is $0.8\,\text{m}^3/\text{s}$ at 7.6 Pa.

Equal friction method

Selected air velocity through duct A = 5 m/s

Calculated airflow through duct A = $0.8\,\text{m}^3/\text{s}$

Calculated airflow through duct B = $0.4\,\text{m}^3/\text{s}$

From design chart Figure 4.36, duct A at $0.8\,\text{m}^3/\text{s}$ = 450 mm dia. with a frictional resistance of 0.65 Pa/m.

Duct B using the same friction at $0.4\,\text{m}^3/\text{s}$ = 350 mm dia. with an air velocity of approximately 4.2 m/s.

The fan will have to overcome $0.65\,\text{Pa} \times 15\,\text{m} = 9.75\,\text{Pa}$. The fan rating is $0.8\,\text{m}^3/\text{s}$ at 9.75 Pa.

The reverse procedure will be more appropriate where limiting ducted air velocity and resistance are paramount to the system design. For instance, if the application is to a library, Table 4.5 shows the need to restrict velocity

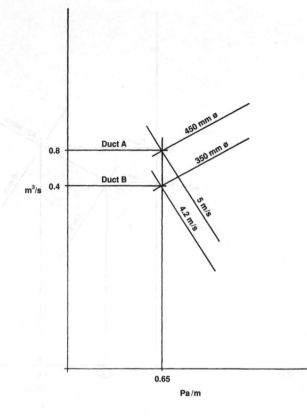

Figure 4.36 Equal friction method

to 2.5 m/s with a maximum resistance of 0.4 Pa/m length of ductwork. From the duct design chart, these two coordinates indicate a maximum discharge of 0.1 m³/s through a 225 mm diameter duct. By calculating the room volume (say 180 m³ for this example), the formula is applied to determine whether sufficient air changes are provided.

$$Q = \frac{\text{Room volume} \times \text{air changes per hour}}{\text{Time in seconds}}$$

$$\text{Air changes per hour} = \frac{Q \times \text{time in seconds}}{\text{Room volume}}$$

$$\text{Air changes per hour} = \frac{0.1 \times 3600}{180}$$

Thus, two air changes per hour would be provided.

5 Air-conditioning principles

The purpose of air-conditioning is to produce and maintain a predetermined internal environment, regardless of external conditions. For human comfort, the objective could be described as provision of a spring-like day all year round. The air temperature should be between 19 and 23 °C and relative humidity within the 40–60 per cent band. To achieve these design criteria, equipment will include facilities to heat, cool, humidify, dehumidify, clean and propel the air in large volumes and velocities. The plant will also have to respond rapidly to atmospheric changes through automatic control.

Systems

Choice of system will depend on building purpose and degree of occupancy. They fall into three categories:

1. All air systems
2. Air–water systems
3. Packaged units.

All air systems

Low velocity

Low-velocity systems of the central plant type shown in Figure 5.1 were originally used to condition large open spaces such as theatres, factories and exhibition centres. The air-handling unit detailed in Figure 5.2 is of the draw-through type with components on the suction side of the fan.

Air-handling unit – function

- Air is usually drawn from the top of the building through a louvred intake.
- Recirculated and fresh air are mixed in the ratio of no more than 3 : 1, i.e. 75 per cent maximum recirculated, 25 per cent minimum fresh air.
- The mixed air is filtered (see Chapter 4).
- In winter, a preheater coil will provide initial heat energy in the air and raise its temperature above zero.
- A water spray washer or steam injector humidifies and cleans the air. The spray temperature may vary, being heated in winter and cooled in summer. Steam humidifiers are now frequently specified in preference to water, to neutralise the possibility of bacteria.

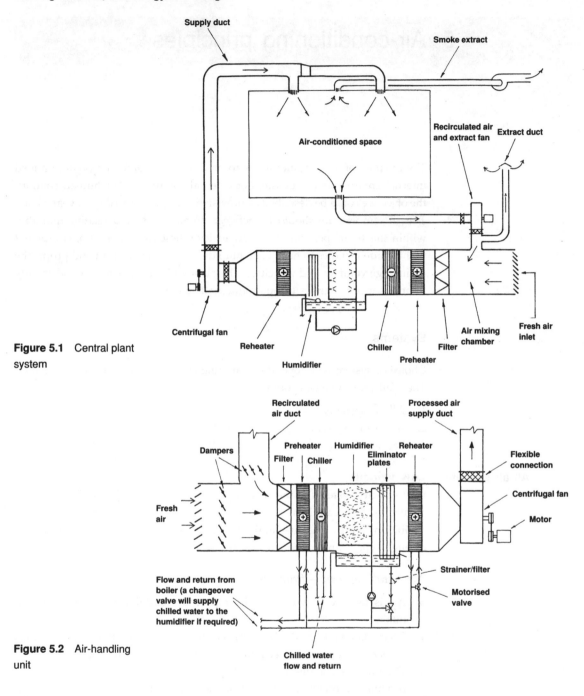

Figure 5.1 Central plant system

Figure 5.2 Air-handling unit

- Eliminator plates remove surplus moisture from the air. These are a bank of corrugated steel or plastic plates which collect excess moisture before recirculation.
- In summer, the air temperature is reduced by a chilled water coil from a refrigeration unit. Efficiency can be improved by using brine as the cooling

medium. Small air-handling units use a direct expansion (DX) coil, i.e. the refrigerant coil or evaporator is suspended in the airstream, to provide direct energy transfer.

- In winter a reheater establishes the supply air temperature and humidity, before fan delivery to the required areas.
- The supply fan, usually of the centrifugal type, generates sufficient pressure and velocity to deliver the air through a system of ductwork to diffusers.

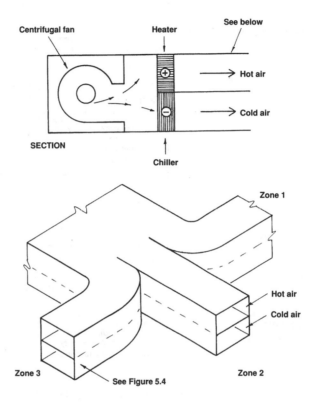

Figure 5.3 Multi-zone air processing

Central plant – zoning

The central plant system has the limitation of the same air quality being delivered throughout the building. If the structure is zoned or divided and compartmented into several different functions such as offices, workshops, canteen, etc. a variety of delivery air properties are possible to suit each situation. The simplest facility incorporates a heating or cooling element in the branch ducts serving particular areas. The temperature of each element can be controlled with zoned thermostats. While adequate for the occasional room, this approach is capital intensive and uneconomic to run if widely deployed throughout a building. A more viable approach would be a multi-zoning arrangement, where the fan precedes the chiller and reheater coils as shown in Figure 5.3, to draw and blow air through the air-handling unit. Several ducts radiate from the unit to designated zones within the building. Each zone receives all hot, all cold or

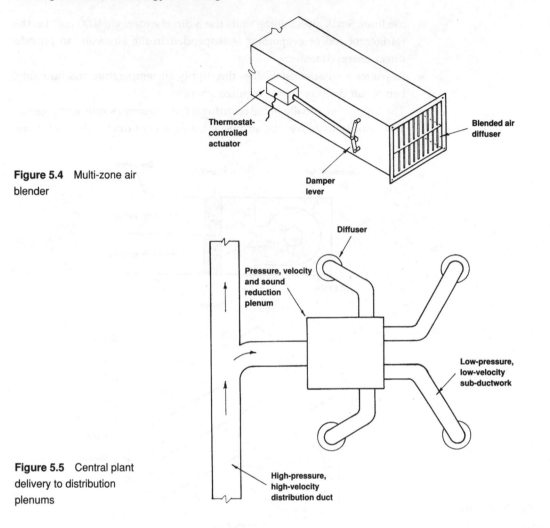

Figure 5.4 Multi-zone air blender

Figure 5.5 Central plant delivery to distribution plenums

a blended intermediate setting, determined by thermostatically controlled motorised dampers shown in Figure 5.4.

High velocity

Central plant
Development of high-rise and large buildings and expectations for sophisticated conditions within these structures, have necessitated considerable improvements to traditional air-conditioning systems. Higher air velocities and pressures are required to overcome distance and restricted accommodation for plant, but it must be remembered that if the ducted design air velocity is increased to reduce the cross-sectional area of ductwork, noise generation will increase. Figure 5.5 shows a simple adaptation from a traditional central plant system. High-velocity air ducting supplies acoustic junction chambers before low-velocity sub-ductwork connects to diffusers. The chamber is lined to absorb

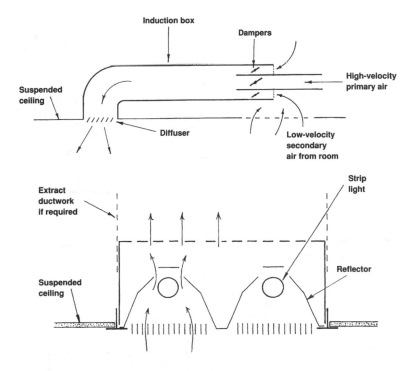

Figure 5.6 Induction diffuser unit

Figure 5.7 Light fitting extract grille

sound and fitted with air volume dampers to regulate final distribution. This system has limited flexibility and if the load or amount of air required is variable due to solar gains, occupancy or intermittent use of machinery, an alternative system should be sought.

Induction units

If the ceiling can be used as a plenum for returning air to the central plant, it is possible to induce a proportion of the room air through a specially made diffuser of the type shown in Figure 5.6. Light fitting extract grilles shown in Figure 5.7 are particularly useful in this situation, having the benefit of utilising the heat from the light, increasing its efficiency and combining ceiling fixtures. A room or zone thermostat will control dampers in the induction box to regulate the proportions of primary and secondary air.

Variable volume

In conventional constant-volume air-conditioning systems, only the air temperature at the air-handling unit is varied. The variable volume system has constant air temperature, but changes the volume of air to suit each room load. Recent research has developed a number of successful terminals, incorporating bellows or motorised dampers of the type shown in Figure 5.8. If the majority of diffusers are closed, a pressure sensor switch could reduce the fan motor output, or open the damper on a bypass duct between supply and return ductwork shown in Figure 5.9. For winter use, it can become an air–water system, with incorporation of a reheat unit in the diffuser. A room

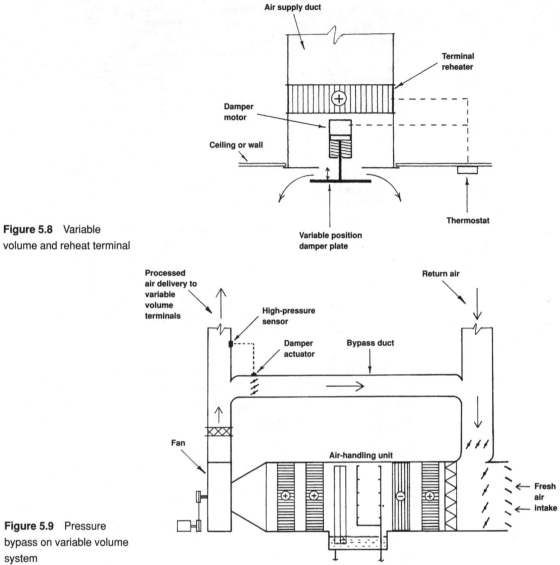

Figure 5.8 Variable volume and reheat terminal

Figure 5.9 Pressure bypass on variable volume system

thermostat will activate a motorised valve on the hot water supply and air damper simultaneously. In summer, only the damper will be active.

Dual duct

The dual duct system is a development of the central plant multi-zoning system, previously described. The preceding system is limited to supplying large areas from air mixed at the air-handling unit, whereas the dual duct system permits individual mixing of cool and warm air at each room. The air-handling unit is without the chiller or reheat coils, as these are deployed in separate ducts branching from the main supply duct, as shown in Figure 5.10. Although the air velocity is high, resulting in relatively small ductwork, substantial space

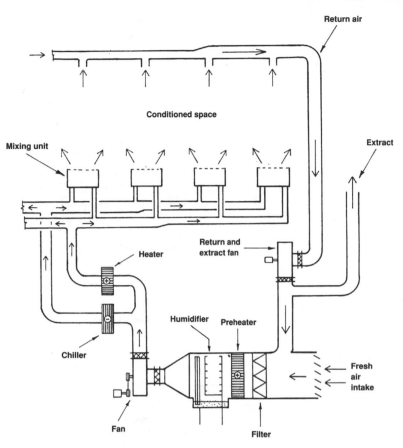

Figure 5.10 High-velocity dual duct system

is needed to accommodate two ducts. However, the simple blending of airs at each terminal, manually or automatically as shown in Figure 5.11, provides a high degree of flexibility. Blending units may be wall or ceiling mounted. It is possible to combine the dual duct system with variable air volume terminals, to attain the advantages of both systems.

Air–water systems

Terminal reheat

This possibility was discussed under limited zoning of central plant all-air systems, where the occasional room has a reheat unit like that shown in Figure 5.12 attached to the branch duct. It is uneconomic to incorporate reheaters before every diffuser in a building, as both capital and running costs would be very high. Having expended energy to cool air and then to incur further expense in reheating it, is not financially viable.

Induction units

Perimeter induction units, so-called because of their popular location beneath windows, provide a blend of primary conditioned air from an air-handling unit to mix with induced secondary air from within each room. The system is pipework intensive, with heating and chilling plant providing energy in

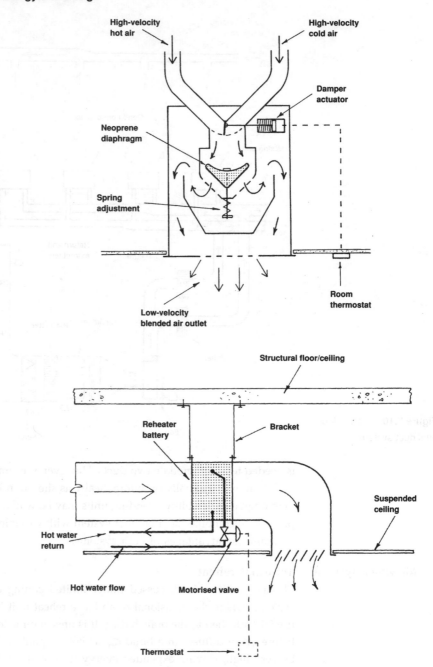

Figure 5.11 Dual duct, air-blending unit

Figure 5.12 Terminal reheat unit

water to the air-handling unit and individual induction units, as shown in Figure 5.13.

High-velocity primary air is propelled through nozzles inside the induction unit, to create a negative pressure in its wake. This draws or induces the secondary room air to blend at the required delivery temperature in a ratio of between 3 and 6 : 1. A discharge or throw of up to 6 m is possible from the

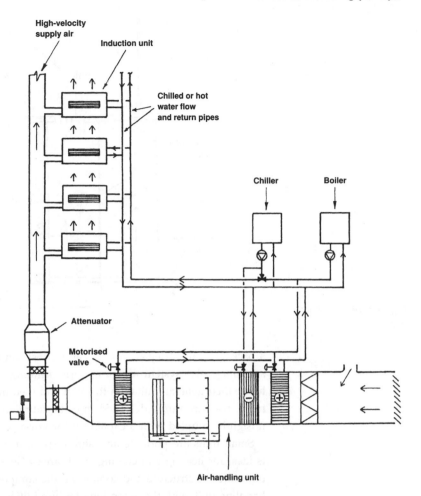

High-velocity supply air

Induction unit

Chilled or hot water flow and return pipes

Chiller

Boiler

Attenuator

Motorised valve

Air-handling unit

Figure 5.13 High-velocity induction unit system *Note*: Return air and extract ducts as Figures 5.1 and 5.10.

units shown in Figure 5.14. The two-pipe 'change-over' system featured in Figure 5.13 is the simplest, with hot water supplying the unit's energy exchange coil for winter use and chilled water for summer use. This economic installation has limited efficiency and effectiveness between seasons, when either heating or cooling could be required. The four-pipe or 'non-change-over' system provides better control, with separate heating and chilling coils inside the unit, but plant and installation expenditure is much higher. A compromise is a three-pipe arrangement, where both hot and chilled water are supplied to the unit and a common return pipe conveys the water back for reheating or refrigeration, depending on its temperature. Beneath the chiller coil is a drip tray to collect condensation for external discharge.

Packaged units   Packaged air-conditioning units are mass produced, factory manufactured to suit a range of applications. This reverses the designer's role of combining individual components to create a purpose-made system to complement a building's characteristics, to adapting these characteristics to match the potential of

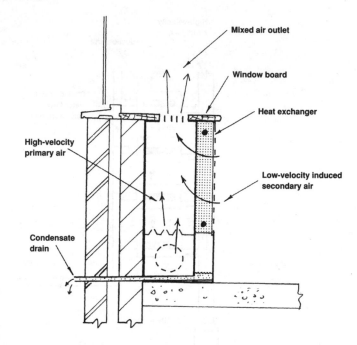

Mixed air outlet

Window board

Heat exchanger

High-velocity primary air

Low-velocity induced secondary air

Condensate drain

Figure 5.14 Induction unit

a packaged unit. They are low in capital cost compared with a fully designed and integrated system, but long-term running costs are likely to be high, hence their general limitation to small and medium-sized structures. Furthermore, they are rarely designed into the fabric of the building and can look unsightly, coupled with a capability for noise generation.

Small units are portable and simply plug in to a three-pin socket, which is ideal for heating and cooling small areas. Generally, the units are fixed or built into the structure and contain all the components of a conventional air-handling unit, with the exception of a humidifier. Nevertheless, humidity can be provided to some extent by the condensation from the chiller. Chilled water coils and associated refrigeration plant, plus cooling towers, are inappropriate for these relatively small items of plant, so incoming air is cooled by DX. This means that the evaporator (cold coil) in a refrigeration or heat-pump cycle shown in Figure 5.15 is directly exposed to the incoming air. The condenser (hot coil) is fan-cooled externally. This is ideal for summer use, and in winter the heat-pump cycle can be reversed as in Figure 5.16, to create a heating coil in direct contact with the incoming air.

Refrigeration

Most refrigeration units used in air-conditioning plant, operate on the vapour compression cycle principle. The refrigerants, known collectively as hydrocarbons and designated HCFCs, have very volatile properties:

- As a fluid, saturation pressure and temperature increase and vice versa.
- During change of state from liquid to gas, the heat absorption is considerable. This is known as the latent heat of vaporisation.

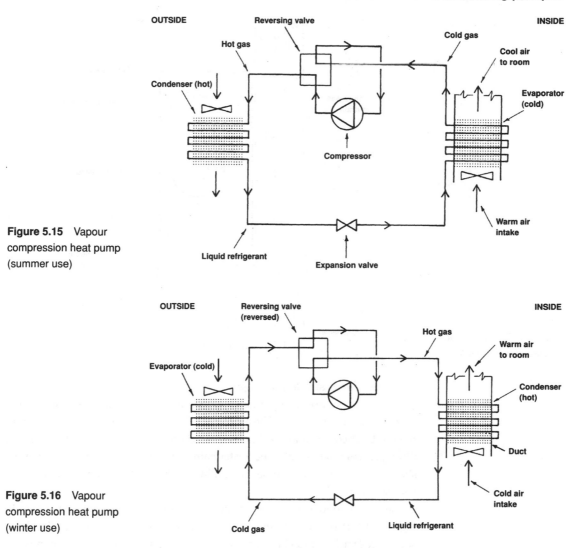

Figure 5.15 Vapour compression heat pump (summer use)

Figure 5.16 Vapour compression heat pump (winter use)

As can be seen in the preceding figures, a pump compresses the refrigerant gas to create a resistance against the regulating valve. This raises the temperature of the gas and changes it to a liquid, emitting the heat energy at the condenser. Conversely, as the liquid escapes through the expansion or regulating valve, it vaporises and absorbs heat energy from the surrounding air at the evaporator coil, manifesting in an ice-cold surface. The cycle continues with the pump recompressing the gas. (*Note:* Refrigerants boil at about –30 °C.)

Types of packaged unit

Packaged air-conditioning systems divide into two main categories:

1. Self-contained, single packages
2. Split packages.

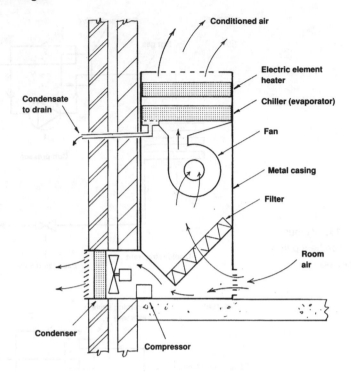

Conditioned air

Electric element heater

Chiller (evaporator)

Condensate to drain

Fan

Metal casing

Filter

Room air

Condenser

Compressor

Figure 5.17 Single duct packaged unit, <6 kW

Self-contained, single package

The simplest example is a portable unit, sometimes described as a single duct unit shown in Figure 5.17. The air to cool the condenser is taken from within the conditioned room and a flexible duct discharges the warm air externally, through a void in the wall. The heater battery at the discharge is usually of the sheathed element electric type, but could be a hot water coil connected to the heating circuit.

The most widely used self-contained package is the window unit shown in Figure 5.18. It is cheap to buy and simple to install, therefore applicable to small areas such as motel rooms and shops, the latter frequently accommodating the unit above the doorway. The disadvantages are the unsightly appearance, fan and compressor noise and limited air distribution.

Single packaged units are not always small and the roof unit shown in Figure 5.19 can be produced in packages up to 300 kW. Supply air is normally connected to a system of ductwork for internal distribution and most units (depending on location), are available with an optional heating coil.

Split packages

Split packages are divided into two components:

1. The fan coil or air-handling unit containing a filter, fan, evaporator (cold) coil and the expansion or regulating valve
2. The condensing unit, containing the condenser (hot) coil, fan and compressor or pump.

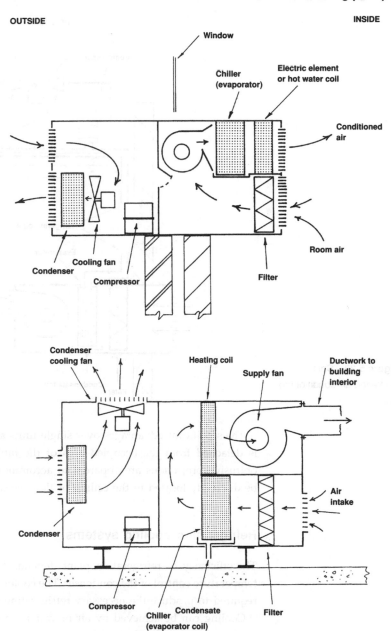

Figure 5.18 Single packaged window unit, <7 kW

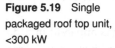

Figure 5.19 Single packaged roof top unit, <300 kW

Figure 5.20 shows the principle, with the two detached casings forming an indoor air-handling unit and an outdoor condensing unit. The common link is the refrigerant pipes; these have a practical limit of about 30 m, although some manufacturers claim greater distances with contemporary refrigerants. One condensing unit can service several air-handling units, increasing the design potential for multi-zone situations, and capacities vary through the same range as for single packages from a few kilowatts to several hundred.

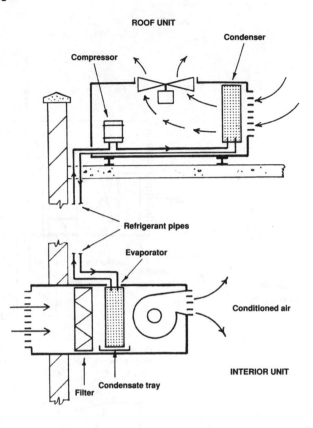

ROOF UNIT

Condenser

Compressor

Refrigerant pipes

Evaporator

Conditioned air

INTERIOR UNIT

Filter

Condensate tray

Figure 5.20 Split packaged air-conditioning unit

The significant advantages over single units are that the noisy compressor is detached from the occupied area of the building and no large holes or obtrusive projections are apparent to accommodate the fan coil unit. It can be discreetly located in the ceiling or floor void.

Refrigeration cooling systems

The efficiency of refrigeration plant depends to a large extent on effective temperature control of the condenser. The cooler the condenser, the less power required to produce the necessary refrigerating effect.

Cooling can be achieved by air or water. The former is apparent in a domestic refrigerator, with the condenser coil exposed to ambient air at the rear of the unit. Packaged air-conditioning systems go one stage further, with fan-assisted cooling of the condenser (see Figures 5.17–5.20), but for large-scale centralised air-conditioning, specific plant is necessary. Until recently, air-cooled condensers had been limited to packaged systems, but numerous outbreaks of Legionnaires' disease in the 1970s and 1980s, subsequently traced to condenser water-cooling towers, has promoted developments in air cooling. Air is a less efficient coolant and to succeed it requires high-powered fans to generate an artificial draught through the condenser coils. Figure 5.21 shows a

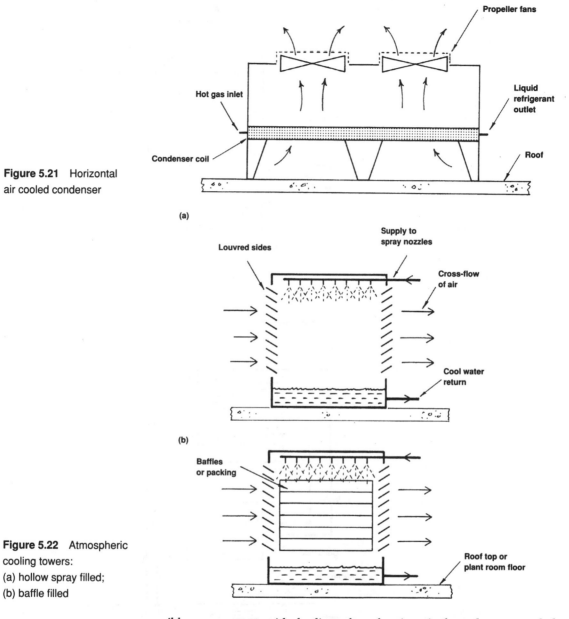

Figure 5.21 Horizontal air cooled condenser

(a)

(b)

Figure 5.22 Atmospheric cooling towers:
(a) hollow spray filled;
(b) baffle filled

possible arrangement, with duplicate fans drawing air through a suspended condenser.

Water cooling of condensers can take several forms, including the use of ponds and ornamental fountains. The most common applications are evaporative coolers of the natural draught or fan-draught types. Natural draught towers are located on the building roof and rely on airflow through their louvred walls, to reduce the temperature of water discharging from a bank of spray nozzles. Plastic baffles can be provided within the tower, to increase the wetted surface and air contact time. Both are shown in Figure 5.22.

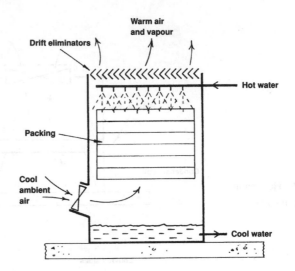

Figure 5.23 Forced-
draught cooling tower

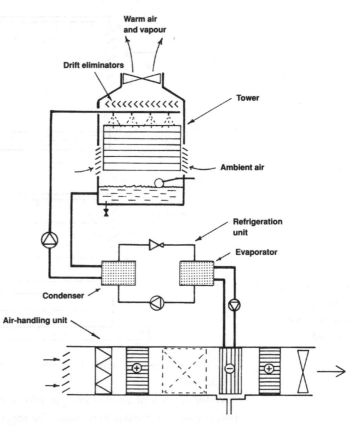

Figure 5.24 Induced-
draught cooling tower

Fan-draught cooling towers are more efficient, being less reliant on the whims of the weather. The forced-draught type shown in Figure 5.23 has a fan at low level which blows ambient air across rows of plastic packing in the face of water droplets form the high-level bank of spray nozzles. Figure 5.24 shows

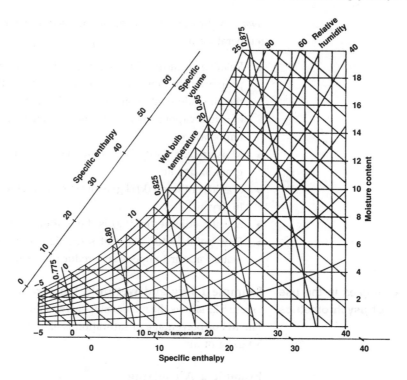

Figure 5.25
Psychrometric chart

the alternative induced-draught tower, complete with refrigeration and air-processing system. Here the air is drawn through the unit from louvred voids near the base. A larger fan permits economy in use and less noise and vibration. (*Note:* In the interests of public health, cooling water must be treated with a biocide. The towers must also be cleaned and disinfected at intervals defined by the Health and Safety Commission affected through the Workplace (Health, Safety and Welfare) Regulations 1992.)

Psychrometrics and air-conditioning design

Psychrometry describes the properties of moist air. Air and water vapour combine to provide different characteristics at different temperatures and these conditions can be represented on a psychrometric chart. The example shown in Figure 5.25 contains information relating to the following:

- Dry bulb temperature (°Cdb). The temperature of air, measured with an ordinary thermometer with a dry sensing element.
- Wet bulb temperature (°Cwb). The temperature of moist air, measured with a thermometer having a damp wick attached to the mercury bulb. If both wet and dry bulb temperatures are the same, the air is 100 per cent relative humidity, i.e. saturated. Normally, wet bulb temperature is less than dry bulb and the difference, which is a measure of humidity, is known as the wet bulb depression.

(*Note:* A sling psychrometer can be used to determine both dry and wet bulb temperatures.)

- Relative humidity (RH) is expressed as a percentage of saturation. It is the proportion of moisture in the air, relative to the amount of moisture that the air could contain at a given volume. It can also be defined as the ratio of actual vapour pressure to the saturated vapour pressure at the same air temperature.
- Moisture content (g/kg) is the mass of water vapour contained in a kilogram of dry air.
- Specific enthalpy (kJ/kg) is the sum total of sensible and latent heat energy in dry air.
- Specific volume (m^3/kg) is the volume occupied by 1 kg of air.
- Dew point is 100 per cent saturation of air and it occurs when air and water vapour temperatures reduce to produce condensation.

Examples in the use of psychrometrics

Example 1 Dry and wet bulb temperatures and RH

Figure 5.26 shows the relationship, given temperatures of 15 °Cdb and 10 °Cwb. The RH or percentage of saturation is 52 per cent and the moisture content is 5.4 g/kg of air.

Example 2 Air mixing

Figure 5.27 shows recirculating ductwork containing air at 21 °Cdb and 15 °Cwb, mixing with fresh air at 36 °Cdb and 25 °Cwb before processing in an air-handling unit. If the ratio of mixed air is 3 : 1, i.e. 75 per cent recirculated to 25 per cent fresh, by plotting both known conditions on the chart and

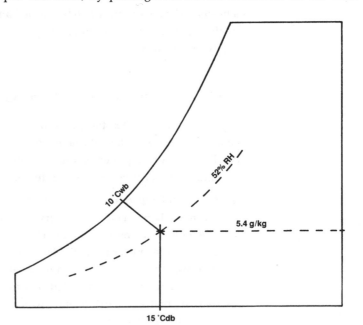

Figure 5.26
Psychrometric applications
(Example 1)

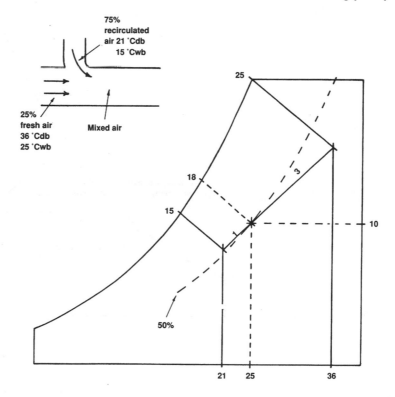

Figure 5.27
Psychrometric applications
(Example 2 air mixing)

proportioning the line between them, the state of mixed air can be obtained. This will be 25 °Cdb, 18 °Cwb, 50 per cent RH and a moisture content of 10 g/kg.

Before considering the chart as a means for plant sizing, it is important to identify how the processes in an air-handling unit are represented. Figure 5.28 shows sensible energy exchange as a horizontal line. Cooling is plotted right to left and heating left to right. Further cooling by saturation follows the 100 per cent RH line downwards. Adiabatic humidification in a spray washer ascends the wet bulb line, until saturation point. Steam humidification of air (now frequently used to eliminate bacteria), is represented by ascending the vertical dry bulb line.

Example 3 Plant sizing

1. An air-conditioning system is used to cool supply air at 27 °Cdb and 20 °Cwb to 20 °Cdb and 14 °Cwb, in a factory of 1500 m³ volume requiring five air changes per hour. The chiller and reheater rating can be determined by a combination of calculations, with data extracted from the process plotted on the chart, shown in outline in Figure 5.29.

$$Q = \frac{\text{Volume} \times \text{air changes per hour}}{3600}$$

$$= \frac{1500 \times 5}{3600} = 2.1 \text{ m}^3/\text{s}$$

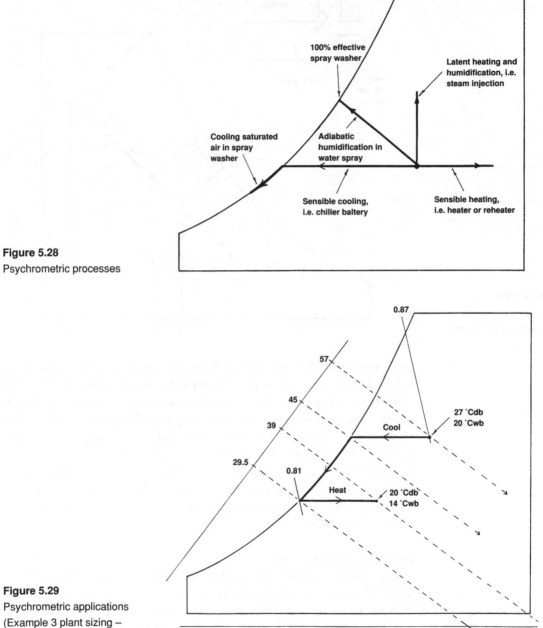

Figure 5.28
Psychrometric processes

Figure 5.29
Psychrometric applications
(Example 3 plant sizing –
summer)

Convert m^3/s to kg/s by establishing specific volume at plant commencing conditions:

$$\text{At } 27\,°\text{Cdb and } 20\,°\text{Cwb} = 0.87\,\text{m}^3/\text{kg} \quad \text{(chiller)}$$

$$\text{At } 10\,°\text{Cdb and } 10\,°\text{Cwb} = 0.81\,\text{m}^3/\text{kg} \quad \text{(reheater)}$$

Therefore

$$\frac{2.1\,\text{m}^3/\text{s}}{0.87\,\text{m}^3/\text{kg}} = 2.4\,\text{kg/s} \quad \text{(chiller)}$$

and

$$\frac{2.1\,\text{m}^3/\text{s}}{0.81\,\text{m}^3/\text{kg}} = 2.6\,\text{kg/s} \quad \text{(reheater)}$$

Enthalpy values for chilling and reheating:

$$\text{Chilling } 57 - 45\,\text{kJ/kg} = 12\,\text{kJ/kg}$$

$$\text{Reheating } 39 - 29.5\,\text{kJ/kg} = 9.5\,\text{kJ/kg}$$

The chiller rating is

$$2.4\,\text{kg/s} \times 12\,\text{kJ/kg} = 28.8\,\text{kW}$$

The reheater rating is

$$2.6\,\text{kg/s} \times 9.5\,\text{kJ/kg} = 24.7\,\text{kW}$$

2. Supply air at –1 °Cdb and –3 °Cwb is required at the same delivery condition of 20 °Cdb and 14 °Cwb in the same factory. By plotting the preheat and adiabatic saturation process on the chart in Figure 5.30, the preheater rating can be calculated.

Convert m^3/s to kg/s at commencing condition:

$$\text{At } –1\,°\text{Cdb and } –3\,°\text{Cwb} = 0.77\,\text{m}^3/\text{kg} \quad \text{(preheater)}$$

Therefore

$$\frac{2.1\,\text{m}^3/\text{s}}{0.77\,\text{m}^3/\text{kg}} = 2.7\,\text{kg/s}$$

Enthalpy values for preheating:

$$28\,\text{kJ/kg} - 4.5\,\text{kJ/kg} = 23.5\,\text{kJ/kg}$$

The preheater rating is

$$2.7\,\text{kg/s} \times 23.5\,\text{kJ/kg} = 63.5\,\text{kW}$$

(*Note:* These calculations assume 100 per cent efficiency of energy exchangers and associated plant. In reality, 70 per cent would be more likely, therefore chiller and heater battery ratings should be multiplied by 100/70.)

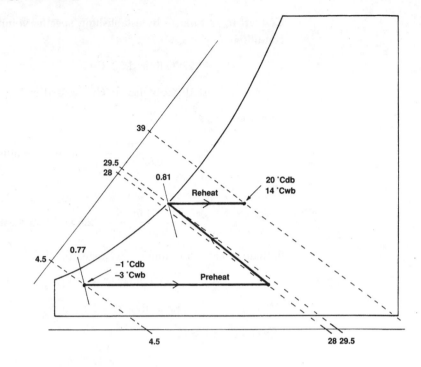

Figure 5.30
Psychrometric applications
(Example 3 plant sizing
– winter)

Health considerations

The growing complexity of modern buildings and the contribution made by
building services, have superficially made contemporary structures more comfortable to use, live and work in. But under the surface, the 'intelligent' building is not universally considered as such. There are serious concerns that the
health and safety of modern building users are threatened. The principal areas
for concern are:

- Legionnaires' disease (*Legionella pneumophila*)
- humidifier fever
- sick building syndrome.

Legionnaires' disease   Legionnaires' disease is a killer. The first notable outbreak was in 1976, during
an American Legionnaires' convention at Philadelphia, USA; 29 people died
out of the 182 who contracted the bacterial infection, and since then there have
been hundreds more incidents throughout the world.

The bacteria thrive naturally in the warm moist conditions of swamps and
similar humid conditions, but have readily adapted to the artificial climate
and environment found in man-made air-conditioning and other service systems. Any permanently moist warm situation, such as dead lengths of pipe and
trapped water in shower roses, could attract the bacteria. The most predomin-
ant breeding area is the packing in the cooling towers of air-conditioning plant,
as detailed in Figure 5.24.

The first major outbreak from this source in the UK killed two and infected nine people at Kingston Hospital in 1980. Since then there have been dozens more, notably at Stafford General Hospital in 1985 where 31 people died, and in 1989 when there was something of an epidemic in the West End of London. The most at risk are the elderly, people with existing respiratory disorders, heavy smokers and those generally in a poor state of health. Nevertheless, there have been several incidences where relatively young, fit and healthy people have been affected.

The bacteria can live and multiply at temperatures between 20 and 60 °C, with optimum reproduction conditions at about 35 °C. Hence the attraction to cooling towers, where they are particularly comfortable in the coarse-faced old timber packing. From here they disperse in water vapour into the atmosphere, some drawn into the air-handling unit and some elsewhere, to expose people within and close to the affected building.

The solution is abolition of wet coolers and replacement with air-cooled condenser units of the type shown in Figure 5.21, or a strict maintenance routine. The latter has been effected through the Health and Safety Commission, and air-conditioned building owners are now expected to:

- undertake regular maintenance of cooling towers, to include cleaning and treatment of the water with biocides
- keep records of maintenance and treatment
- appoint a person or facilities management company to be responsible for supervision of maintenance.

Furthermore, there is an attempt to form a compulsory register of the estimated 38 000 wet towers in the UK.

Humidifier fever
Humidifier fever is an allergy causing temporary discomfort, with symptoms similar to influenza, i.e. headaches, shivering and general aches and pains. The source has been found in the water reservoirs of humidifiers shown in Figure 5.31, where micro-organisms of the amoebae species breed while the plant is shut down for weekends and holidays. The dead husks of the amoebae are drawn into the airstream and dry into a fine dust, which is inhaled by the building occupants. Biocidal treatment to the spray water is one solution, or replacement with a steam injection humidifier.

Sick building syndrome
Sick building syndrome may manifest itself as any one of a number of symptoms, or a combination or collection of ailments such as headaches, lethargy, skin irritations, dry or running eyes, runny nose, throat inflammation and loss of concentration. These occurrences are notably more prevalent in contemporary buildings, which are sealed and have artificially controlled environments.

Sick building syndrome is therefore something of a mystery, defined in symptoms but not causes. Fortunately it is not deadly or disabling, but a sufficient contributor to absenteeism from work to promote considerable research. Suspect areas for investigation include:

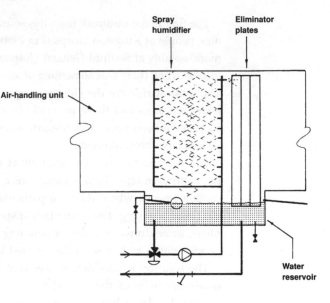

Figure 5.31 Water spray air humidifier

- noise, e.g. background humming from air movement in ductwork and buzzing from fluorescent lights
- artificial lighting, strobe effect of fluorescent fittings
- static electricity from photocopiers, computer monitors, etc.
- glare from computer monitors and light fittings
- psychological disorientation caused by tinted windows, no windows and lack of visual contact with the outside world
- insufficient privacy of the modern office environment
- lack of direct control of the internal environment, i.e. intelligent, fully automated building.

There is little evidence to support any one single cause and although the problems are evident in modern artificially controlled environments, air-conditioning itself should not be maligned. Sick building syndrome has also been recognised in more traditionally serviced buildings, albeit on a much smaller scale.

Air pollutants are closely related to sick building syndrome and suspect sources include poorly maintained ventilation and air-conditioning systems, as well as misuse. This includes minimising or even eliminating fresh air from the system, by recirculating the same air continuously. While economising in fuel, it is likely to produce a stale environment reducing human output considerably. Such abuse could attract legal proceedings under the Health and Safety at Work, etc. Act and the Workplace (Health, Safety and Welfare) Regulations.

Other suspect sources are cleaning equipment and materials, plus synthetic fibres/materials used in modern carpets and soft furnishings. Organic dust pollution and microbiological infestation from dust or carpet mites is well charted as a possibility.

In reality, different people are likely to respond to different atmospheres. Some will respond to a cocktail of contaminants and others not. There is therefore no conclusion to this subject, but it does question why we do not suffer from sick house syndrome. Perhaps that will be the next mystery ailment, as the current trend is to construct more tightly sealed and highly insulated dwellings!

Energy recovery

The known energy resources of the world are diminishing and conventional fuels cause atmospheric pollution, the greenhouse effect and contribute to ozone depletion. It is, therefore, of paramount importance that energy be reused where possible. High tariffs and taxation alone are insufficient deterrents; science and technology have to combine to develop more efficient fuel-consuming equipment. In addition to the condensing boiler, which reuses the flue gases for secondary heating (see Chapter 3), some other successes include:

- heat pump
- heat pipes
- plate or annular heat exchanger
- run-around coil
- thermal wheel.

Heat pump The components and operating principles of a heat pump are similar to the vapour compression cycle of refrigeration, considered previously in this chapter. With refrigeration, the evaporator cools and heat is rejected at the condenser. Considered in reverse, with the evaporator drawing heat from a low-grade heat source such as air-conditioning extract ducts, ventilation extracts and chimney flue gases, a high grade or temperature output can be attained at the condenser. In this format, the equipment is known as a heat pump and Figure 5.32 shows a typical unit in operating mode.

Heat-pump manufacturers quote the effectiveness of their products as a 'coefficient of performance' (COP), or as an 'energy efficient ratio' (EER). They are a measurement of performance based on temperature in degrees Kelvin, of the thermodynamic cycle, i.e.

$$COP = \frac{T_c}{T_c - T_e}$$

where T_c = condensing temperature and T_e = evaporating temperature. For example, if T_c = 50 °C and T_e = 5 °C,

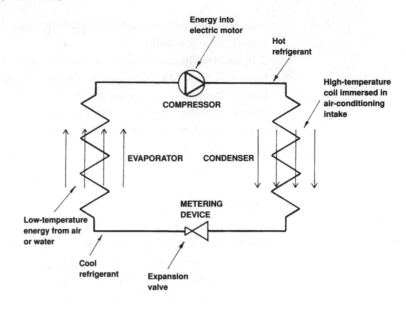

Figure 5.32 Heat pump cycle

$$COP = \frac{50 + 273}{(50 + 273) - (5 + 273)} = 7.17$$

This means that for every 1 kW of energy used in the electrical compressor, a heat output of over 7 kW is possible. In reality a COP of about 3 is more likely. This takes into account the efficiency of the compressor (which decreases with age), the application, the weather/time of year and the effectiveness of other components and the installation.

Heat pipes   These consist of a battery or bank of parallel copper tubes, mounted between adjacent air intake and extract ductwork in ventilation or air-conditioning systems, as shown in Figure 5.33. A refrigerant is sealed into each tube and if used horizontally, a wick lines the tube to encourage the working fluid to return to the evaporator after cooling. If mounted vertically, the wick is unnecessary as the process will occur by gravitation.

The heat pipe is divided into two parts (assuming equal mass flow rates between intake and extract), with one end immersed in the cold air intake and the other in the warm air extract. The warm air causes the fluid to evaporate and migrate to the cooler end of the pipe, due to the changes in vapour pressure. Here the vapour condenses into a liquid and the change of state heat energy transfers to the surrounding cool air. Capillary action in the wick encourages the fluid to return to the evaporator. The concept is simple and maintenance free (apart from cleaning) as there are no moving parts.

Given a heat pipe installation with an efficiency of 60 per cent, immersed in equal air flow rates, the following calculation shows how to determine the recovered supply air temperature (T_s) if outside air (T_o) is –1 °C and extracted stale air (T_e) is 20 °C:

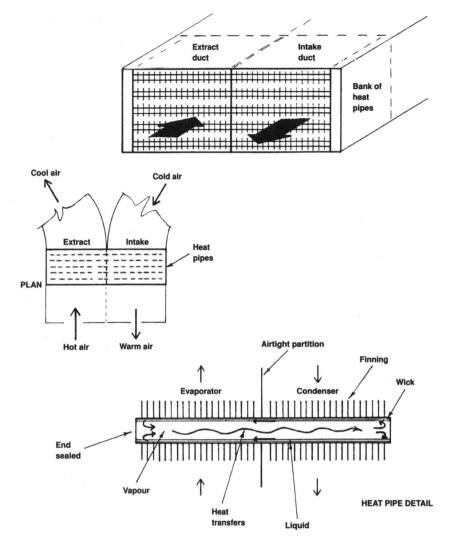

Figure 5.33 Heat pipes

$$T_s = T_o + [\text{efficiency}(T_e - T_o)]$$

$$= -1 + [0.6(20 - -1)]$$

$$= 11.6\,°C$$

Plate or annular heat exchanger

These are the simplest type of energy recovery unit. They have the warm exhaust air from an air-conditioning or ventilation system counterflowing through the cold intake air. The plate type is compartmented, with the airstreams conducted through a series of separating plates as shown in Figure 5.34. Also shown is the simpler annular arrangement of warm extract ductwork inside the cold air intake. Efficiencies of over 50 per cent are claimed, but will depend considerably on air speeds.

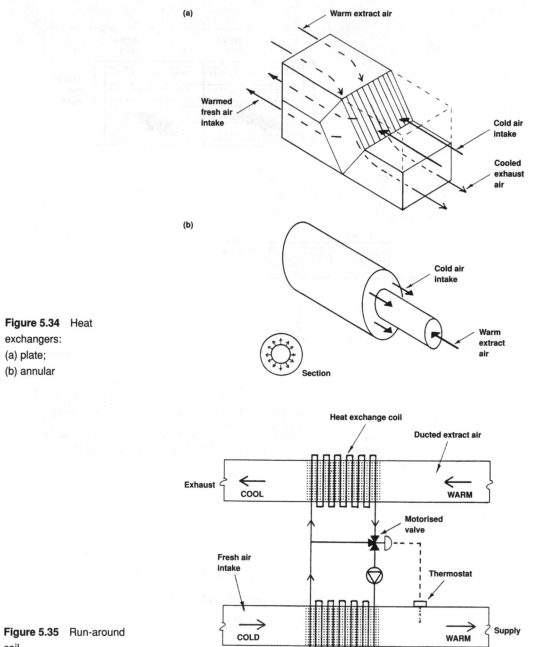

Figure 5.34 Heat exchangers:
(a) plate;
(b) annular

Figure 5.35 Run-around coil

Run-around coil This system is based on two finned tube energy exchangers. One is immersed in warm extract air from ventilation and air-conditioning systems or flue gases, the other in the cold air intake to ventilation or air-conditioning. The two coils are linked with a piped circuit of pumped water and glycol (anti-freeze) solution as shown in Figure 5.35. The exhaust air energy exchanger collects and

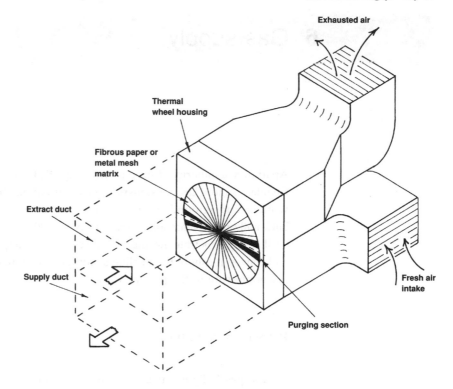

Exhausted air

Thermal
wheel housing

Fibrous paper or
metal mesh
matrix

Extract duct

Supply duct

Fresh air
intake

Purging section

Figure 5.36 Thermal wheel

transfers the heat through the water to the other energy exchanger in the air intake. It can be applied where heat recovery and transfer are fairly remote, but efficiency is unlikely to exceed 40 per cent due to long pipe runs and limited thermal properties of water.

Thermal wheel

This is otherwise known as a regenerative heat exchanger or energy recovery wheel. The large wheel of up to 4 m diameter shown in Figure 5.36, circulates through adjacent exhaust and intake airstreams. The concept originated over half a century ago and proved successful with a core of metal mesh. Subsequently, research and development have produced aluminium and inorganic fibrous paper cores coated or impregnated with lithium chloride. The lithium chloride is particularly effective at absorbing latent heat energy from the water vapour in warm stale exhaust air from air-conditioning.

Rotation is only about 10–20 rpm, with half the wheel in contact with warm exhaust air and the other half in contact with cool intake air. Effective energy transfer occurs because a labyrinth of seals in the purging section prevent cross-contamination of airs. Considerable space is required for these wheels, but claims of 90 per cent effectiveness are made by some manufacturers.

6 Gas supply

Apart from a minority of supplies from liquefied petroleum gas, considered in Chapter 3, most UK consumers receive natural gas from a distribution network of underground mains. The origin is decayed organic matter, dating back millions of years and found at up to 3000 m below the North Sea bed. Drilling rigs penetrate the impervious strata which traps the gas, tap the source and convey it ashore by pipeline to compressor stations which maintain sufficient pressure in the pipe network.

Properties of natural gas

Natural gas is lighter than air, having a relative density of about 0.55. Its composition is essentially hydrocarbon, with a nominal amount of oxygen and nitrogen as defined in Table 6.1. In order for combustion to occur, there must be sufficient heat and oxygen. Ignition temperature is just over 700 °C and oxygen for combustion is required at twice the volume of methane. Therefore, as air contains about 20 per cent oxygen, approximately 10 per cent gas to air mixture will be needed to satisfy combustion.

$$CH_4 + 2O_2 \longrightarrow CO_2 + 2H_2O$$

i.e. methane (1 part) + oxygen (2 parts) produces carbon dioxide (1 part) and water vapour (2 parts).

Table 6.1 Approximate composition of natural gas

Constituent	Chemical formula	Percentage composition
Methane	CH_4	90
Ethane	C_2H_6	5
Propane	C_3H_8	1.5
Butane	C_4H_{10}	0.5
Carbon dioxide	CO_2	0.5
Nitrogen	N_2	2.5

Service connections

A network of gas mains service most towns and cities in the UK. Much of this is very dated steel pipework, wrapped in a protective grease tape. British Gas PLC are continually upgrading these with replacements in distinctive yellow uPVC and polyethylene branch supplies to individual buildings. Prospective consumers must apply to the authority for a connection and charges are usually negotiated on a fee per metre run. Some of the cost may be absorbed if the gas authority recognises the business potential of supplying additional buildings. Installation must comply with the safety requirements defined in the Gas Safety (Installation and Use) Regulations and Part J to the Building Regulations. All gas operatives and approved contractors will be registered with CORGI, the Council for Registered Gas Installers.

Installation

Service pipes will vary in diameter depending on demand from commercial and individual premises. A method for pipe sizing concludes this chapter, but for typical domestic consumption a 25 mm bore pipe is quite adequate. Table 6.2 provides guidance for multiple occupancies.

Individual domestic main connection is without an isolating valve, as mains pressures are relatively low, rarely exceeding 75 mbar (750 mm water gauge (w.g.) or 7500 Pa). Supplies over 50 mm in diameter are provided with an accessible service valve where the pipe enters the property boundary. This must be easily identified for closure in event of a fire. Ground cover of 375 mm (450 mm in public places) is sufficient as shown in the typical installation of Figure 6.1. The primary pipework is terminated at a meter before secondary distribution in the building. Meters are preferably located in a purpose-made external box of the type shown in Figure 6.2. They may be:

- semi-concealed
- surface mounted or
- sunken.

The inside of an outside wall is acceptable, if the building has a lobby area serving a block of flats or similar collection of units. Within a garage has been permitted, but current policy is to avoid this if possible, as there is potential

Table 6.2 Service pipe sizes for flats

Nominal bore of pipe (mm)	No. of dwellings/meters
32	2–3
38	4–6
50	Over 6

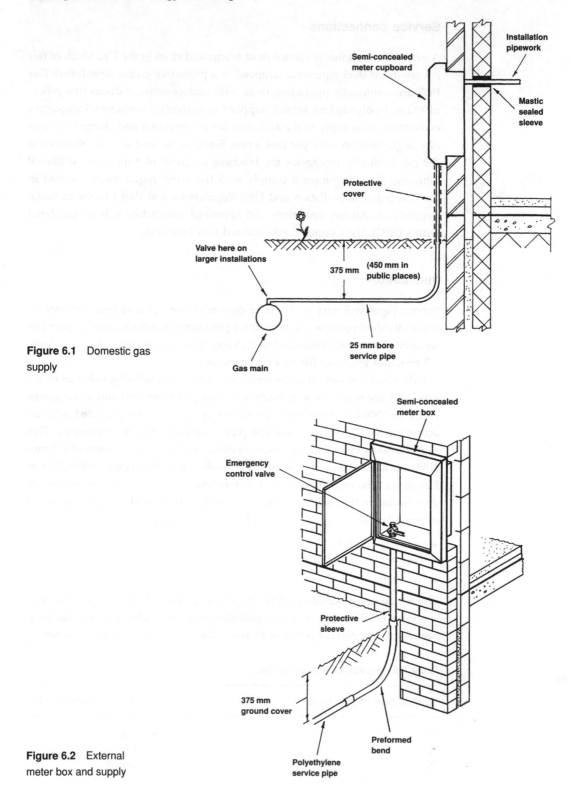

Figure 6.1 Domestic gas supply

Figure 6.2 External meter box and supply

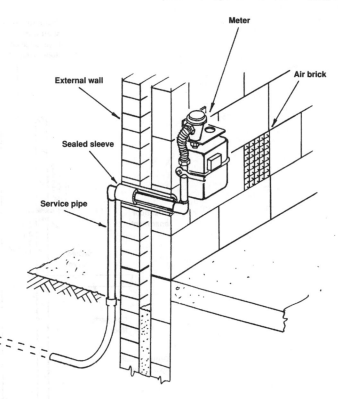

Figure 6.3 Gas meter in ventilated lobby

for damage, dampness and exposure to sparks from workshop activities. The reasoning behind external locations is for:

- ease of periodic reading, without disturbing the building occupants
- less opportunity for bogus officials to gain access to premises
- guaranteed meter access at any time, preventing accounting arrears and fraudulent misuse
- easy emergency access, such as the need for isolation in event of a fire.

To encourage the use of external meters, the gas authority provides the plastic box free of charge to builders, for installation during construction. The colour is normally white, but red/brown may also be available to match brickwork. They may be painted to match the exterior finishes to the building.

The service entry must be sleeved or be provided with a lintel where it passes through a wall, to allow protection against movement. It should be of steel or copper above ground and must not be routed under foundations or through unventilated voids such as wall cavities. Taping of electrical cables to the pipe and location near heat sources are also unacceptable.

Internal meters The traditional practice of siting meters under the stairway is now avoided, but under special circumstances it may still be permitted if the void is well ventilated and the surrounding construction has at least half an hour fire resistance. Figure 6.3 shows the installation in a lobby or utility compartment and

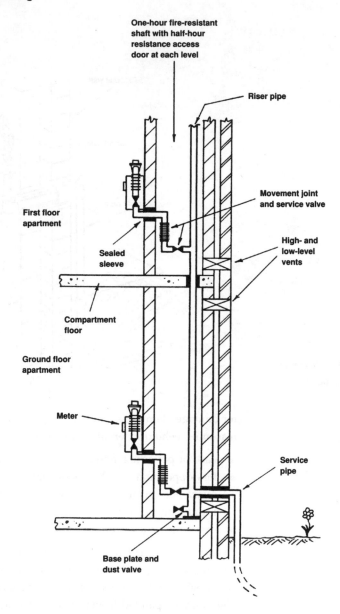

One-hour fire-resistant
shaft with half-hour
resistance access
door at each level

Riser pipe

Movement joint
and service valve

First floor
apartment

High- and
low-level
vents

Sealed
sleeve

Compartment
floor

Ground floor
apartment

Meter

Service
pipe

Base plate and
dust valve

Figure 6.4 Supply to
high-rise units

Figure 6.4 indicates the application to multi-storey properties, where a ventilated service shaft houses an internal gas riser pipe to meters in each dwelling. Safety features include through ventilation air bricks at the top and bottom of the shaft, flexible pipe joints at every branch and fire stopping where the pipe penetrates the structure. Depending on the size and purpose of the building, shafts may also need compartmenting as defined in Part B of the Building Regulations.

Meters Meters remain the property of the gas authority and are supplied with a pressure governor and flexible stainless steel connections as shown in Figure 6.5.

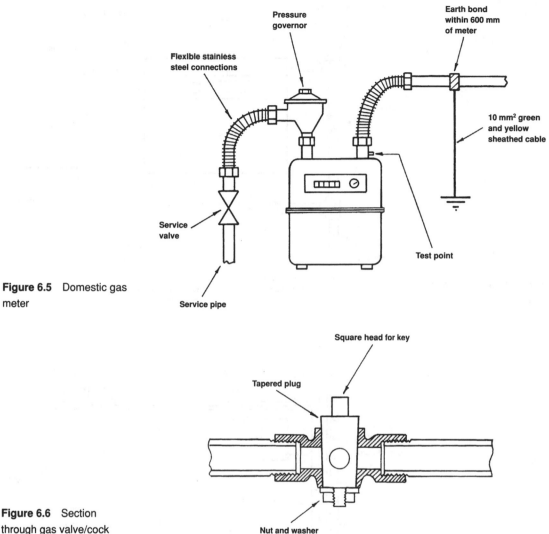

Figure 6.5 Domestic gas meter

Figure 6.6 Section through gas valve/cock

Governors are preset to provide a constant supply at a pressure of 20 mbar (200 mm w.g.) and are lead sealed to prevent unqualified tampering. A gas plug cock precedes the meter to provide means of isolation for maintenance and emergencies. These turn through 90° for simple on/off provision. Figure 6.6 shows the principle of a tapered plug with hole midway, providing passage or not for the low-pressure gas. On the meter outlet is a testing point or nipple for checking supply pressure and system leakage. There is also an electrical equipotential earth bond within 600 mm of the meter, terminating at an earthed rod. This is necessary as an electrical fault within the boiler or its accessories could conduct through the gas pipework.

Meters for industrial or commercial premises have flanged connections to suit 75, 100, 150 and 200 mm nominal bore steel pipes, as shown in Figure 6.7.

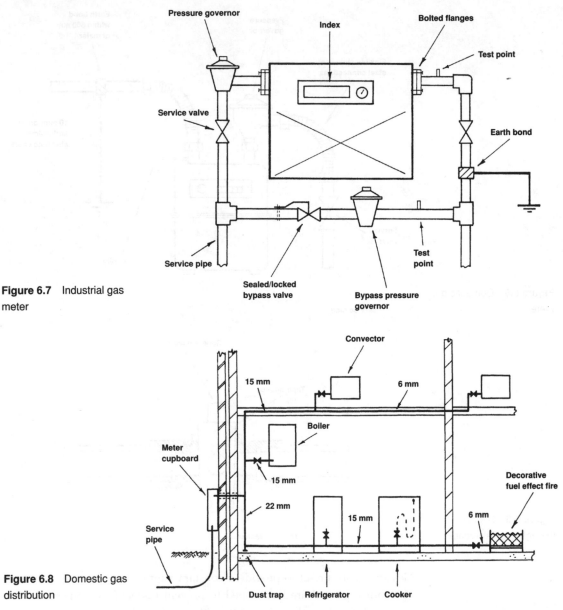

Figure 6.7 Industrial gas meter

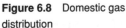

Figure 6.8 Domestic gas distribution

Connections and accessories are similar to domestic meters, excepting flexible couplings which are unnecessary due to the strength of pipe, meter and a firm, level base. With gas authority approval, a bypass pipe and sealed valve are permitted to provide supply continuity if the meter malfunctions, needs repair or maintenance.

Secondary pipework   Internal distribution may be in mild steel pipe or copper tube, the latter preferred for domestic installations. A typical scheme is shown in Figure 6.8. In blocks of flats or hostels, it may be necessary to install several secondary

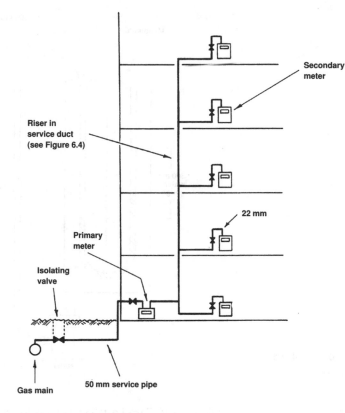

Figure 6.9 Gas supply to flats

meters as shown in Figure 6.9. This is appropriate where apartments are for letting and the landlord retains overall responsibility for gas consumption. Pre-payment type meters could be used, but these have been largely replaced with credit meters.

Purging and testing

Before commissioning the appliances, it is necessary to:

- purge the system of air
- test for soundness.

Purging
The purging process is to remove air, debris, dust and metal filings from the pipework. During this operation good ventilation is essential, as is an absence of naked flames, smoking, use of electric switches, power tools and appliances. Procedure:

1. Close the plug cock before the meter and disconnect secondary pipework at the furthest fitting.
2. Open the control cock and allow at least five times the volume of gas per revolution through the meter. For example, if the meter has 0.07 ft^3/rev,

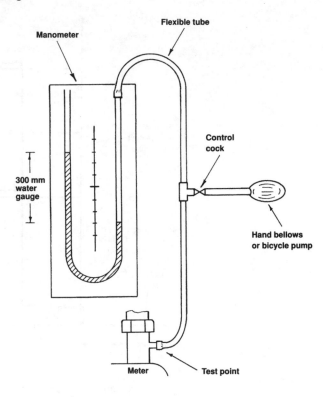

Figure 6.10 Installation
test equipment

allow 5×0.07 ft^3, i.e. 0.35 ft^3 of gas to purge the installation. The smell of gas will be apparent at the disconnected joint.
3. Further branches of pipework need only be opened until gas is smelt.

Testing

Two test procedures are defined, depending on whether the system is (a) new or (b) an existing installation. New installations require all open ends to be capped and fitted appliances turned off at their control valves. If the meter is fitted, the adjacent control cock is closed and the screw removed from the test point. In the absence of the meter, the test is made on the connecting pipe. Figure 6.10 shows the testing apparatus, consisting of manometer (glass U-tube), hand bellows, flexible tubing and test cock.

With the tubing attached to the test point and water levelled at zero in the manometer, the hand bellows are pumped to raise a displacement of 300 mm w.g. The test cock is closed and variations in pressure due to temperature adjustment are allowed for 1 minute. Thereafter for 2 minutes the pressure will be static in a leak-free installation.

Existing installations should undergo two tests:

1. To ensure effectiveness of the main control cock at the meter: all appliance valves are closed as well as the main control cock. The screw is removed from the test nipple and with manometer attached, the control cock is

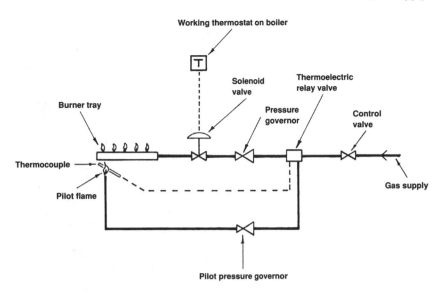

Figure 6.11 Gas burner controls and safety features

opened slightly to record a few millimetres water gauge. The cock is immediately closed and if the pressure rises, the valve is faulty and the gas authority must be informed that it needs replacement.

2. If the cock is serviceable, the test continues to displace between 200 and 250 mm w.g. This is normal domestic appliance working pressure and if the gauge displays anything else, the authority should be informed that the pressure governor is faulty. Assuming correct pressure, the gas is turned off and 1 minute allowed for stabilisation and effects of temperature. After a further 2 minutes, there should be no further pressure reduction.

Leakage detection

If the test proves satisfactory the appliances may be commissioned, but if leakage occurs this must be located and repaired. An obvious smell may indicate the source, otherwise all pipe joints and connections should be sprayed with a leak detection aerosol or brushed with soap solution. A profusion of bubbles will indicate the leak.

Safety controls

Boilers and other heat-producing appliances, which have automatic gas controls, must be fitted with a safety device to prevent the discharge of uncombusted gas if the pilot light fails. The pilot flame failure mechanism shown in Figure 6.11 has a thermocouple suspended in the pilot flame. While hot, the thermocouple energises or opens an electromagnet or solenoid valve (thermoelectric valve) to permit gas flow to the burner. The gas will burn when the boiler working thermostat calls for heat, and when the system is programmed

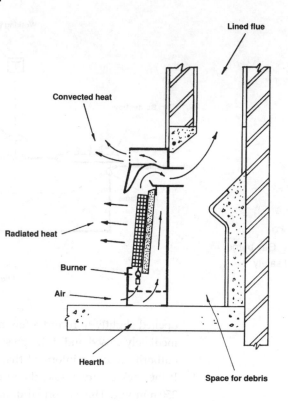

Figure 6.12 Domestic gas fire <3 kW

to function. The thermoelectric valve is initially operated manually with a push button and spark igniter for the pilot light. After 20–30 seconds the button can be released as the electromagnet has energised the valve. If the pilot persistently extinguishes, the thermocouple is most likely in need of replacement.

Appliances

Fires Gas-burning fires are rated up to about 3 kW, providing heat by radiation and convection. Although the capital costs are low, it would be uneconomic to use them throughout a building. They are particularly useful to heat the occasional room or supplement a centralised heat source. Figures 6.12 and 6.13 show application to an existing chimney flue, with air for combustion taken from the room. Purpose-provided ventilation is not normally required for fires up to 7 kW input rating. Popular variations include decorative fuel effect (logs and coal) gas burners shown in Figure 6.14. These are set in traditional solid fuel fire recesses and will require permanent ventilation of at least 100 cm² for appliances up to 15 kW input. This can be through a purpose-made void, with metal vent or perforated clay brick also shown in Figure 6.14. Ventilation is essential with these decorative fuel effect fires, as modern housing is well sealed and insulated against natural draughts.

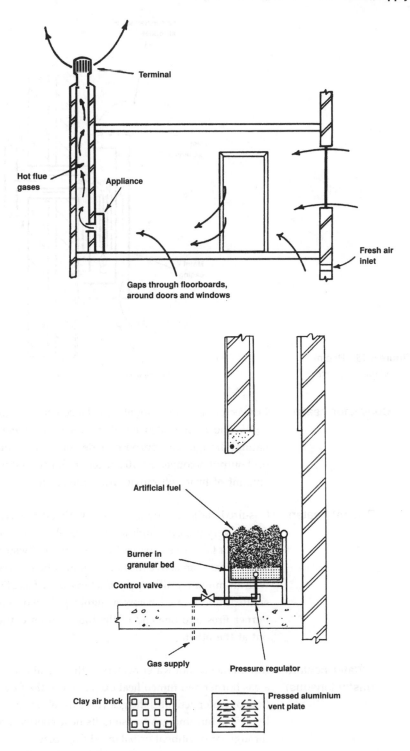

Figure 6.13 Natural draughts to gas appliance

Figure 6.14 Decorative fuel effect gas appliance. *Note:* Ventilation at least 100 cm² for appliances up to 15 kW

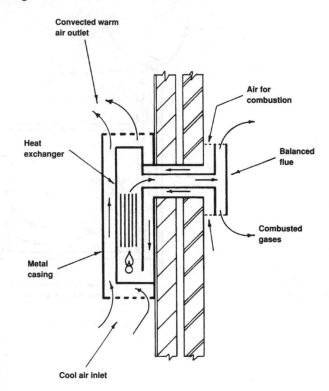

Figure 6.15 Room-sealed gas convector

Convector heaters   Convectors are wall-mounted, balanced flue room heaters rated up to about 7 kW. The room-sealed installation shown in Figure 6.15 is independent of natural draught and purpose-made ventilation voids. Up to 90 per cent of the heat output is convected, 10 per cent radiated, but if fan-assisted only a nominal amount of heat will radiate from the casing.

Radiant heaters   Gas-fired radiant tubular heaters with output typically 10–40 kW, have become a very popular application in warehouses, workshops, factories, churches and other large open spaces. They are generally suspended from the roof structure, at least 3 m above the floor, but may be incorporated behind false ceilings with aluminium grille registers set in the surface. Figure 6.16 shows the radiant emitter tube located below a highly polished stainless steel reflector. A gas burner fires into one end of the tube and an extract fan draws the hot gases out at the other.

Water heaters (instantaneous)   Instantaneous hot water heaters with several draw-offs or 'multi-points', have a gas burner and finned heat exchanger in the flue. As a hot tap is opened, the flow of water activates the gas valve to release gas for combustion by the pilot light. The burning gas transfers its heat energy into the water and discharges through conventional or balanced flue. Sizes vary from single outlet over the sink-mounted heaters, to the multi-point application shown in Figure 6.17 and combination boilers discussed in Chapter 3.

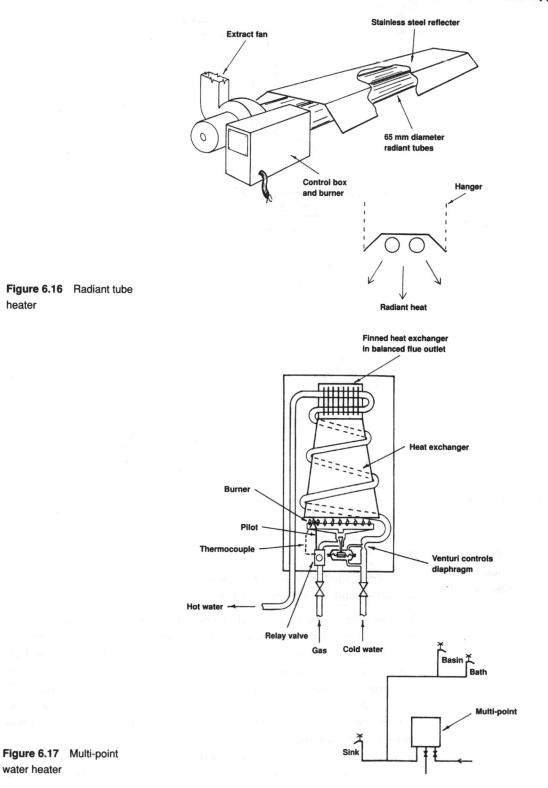

Figure 6.16 Radiant tube heater

Figure 6.17 Multi-point water heater

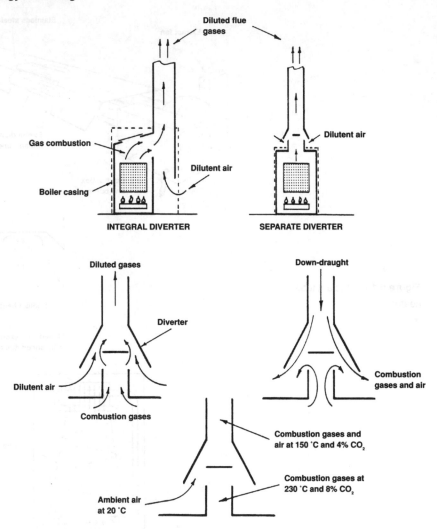

Figure 6.18 Draught diverters

Water heaters and flue systems (traditional boiler)

Boilers are either:

- floor-standing with conventional open flue, or
- wall-mounted with balanced flue.

Floor-standing boilers have conventional open flues incorporating draught diversion to prevent down-draughts extinguishing the pilot light. The diverter also draws in room air to dilute the combustion gases and lower their temperature. Figure 6.18 shows two applications with the effect of diluting air. Air for combustion must be provided by purpose-made voids in the external wall, if the boiler exceeds 7 kW input rating. The accepted rule is 450 mm² of air vent for every 1 kW boiler rating over 7 kW. A 225 mm air brick will satisfy most domestic situations.

Wall-mounted boilers are room sealed and have balanced flues, i.e. the air for combustion is drawn directly into the boiler from outside and the combusted

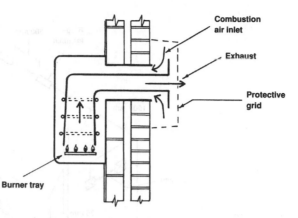

Figure 6.19 Balanced flue

gases discharge through a combined flue/intake unit shown in Figure 6.19. The air inlet and flue gas outlet are adjacent, maintaining constant air pressure through the flue whatever the wind conditions. Fan-assisted balanced flues can be used to increase efficiency of air and combustion gas movement. This reduces the flue size and allows short sections of horizontal flue duct to be fitted if the boiler cannot be conveniently located directly on an external wall. A fan failure device is fitted to prevent the boiler firing if the fan becomes defective.

Balanced flue location

Flue terminal location is important to maintain safety and efficient boiler performance. Table 6.3 lists accepted positions which are complemented by Figure 6.20.

Table 6.3 Minimum dimensions for location of balanced flue terminals

Position	Minimum distance (mm)	
	Natural draught	Fanned draught
Under an openable window or ventilator	300	300
Under rainwater goods or sanitation pipework	300	75
Under eaves	300	200
Under a balcony or a car port roof	600	200
From a window or door in the car port	1200	1200
Horizontally from vertical drain and soil pipes	75	75
Horizontally from internal or external corners	600	300
Above ground level or a projection, e.g. balcony	300	300
From an opposing surface	600	600
From an opposing terminal	600	1200
Vertically from another terminal	1500	1500
Horizontally from another terminal	300	300

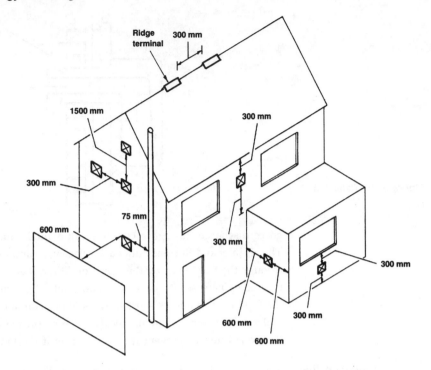

Figure 6.20 Balanced flue and ridge terminal positions (see Table 6.3 for fan-assisted flues)

Boiler rating

Boilers are rated by input and output. The energy input is the amount of fuel consumed and the energy output, the energy in the water. The difference is the energy lost in converting the gas to heat and transferring the heat into the water. For example, if a gas boiler manufacturer states that one of his products has an input of 20 kW and output of 15 kW, the efficiency is

$$\frac{15}{20} \times \frac{100}{1} = 75 \text{ per cent}$$

This figure is typical for a conventional boiler when new, but condensing boilers (see Chapter 3) have efficiencies over 90 per cent.

Other flue systems

Conventional open flue

These have been largely superseded by the simpler balanced flues for domestic boilers, but are still used for higher-rated boilers in industrial and commercial buildings and where it is necessary to site the boiler away from an external wall. They incorporate a draught diverter (or canopy for a gas fire) and an air opening from the room into the gas combustion chamber. The room containing the boiler must therefore be ventilated.

The installation must take account of the terminal location, the flue length and its size. Figure 6.21 shows the most convenient straight flue, with non-combustible sleeve where it passes through the floor and other exposed parts of the structure. Flue height considerations are a balance between gaining draught and a high efflux velocity, with increases in surface friction. Height

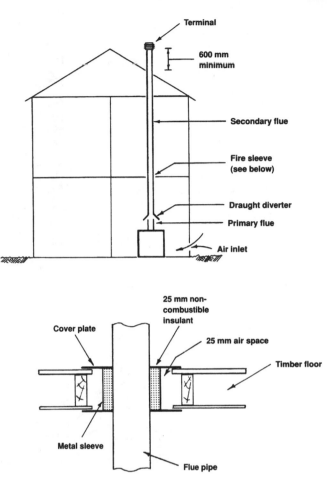

Figure 6.21 Internal open flue

will also increase surface area for heat losses, encouraging condensation in the flue. This occurs at approximately 60 °C, when the gases cool to the dew point of water.

Flue materials and condensation

The most commonly used materials for individual flues are:

- traditional brick chimneys
- precast concrete flue blocks
- flue pipes.

Traditional brick chimneys

These are unnecessarily large for most gas-fired units, allowing flue gases to cool and develop into condensation. New chimneys are lined in accordance with Part J to the Building Regulations and existing unlined chimneys should be fitted with a stainless steel flexible liner, as detailed in Figure 6.22. No lining will encourage the combination of moisture and combustion products to break down the old mortar joints.

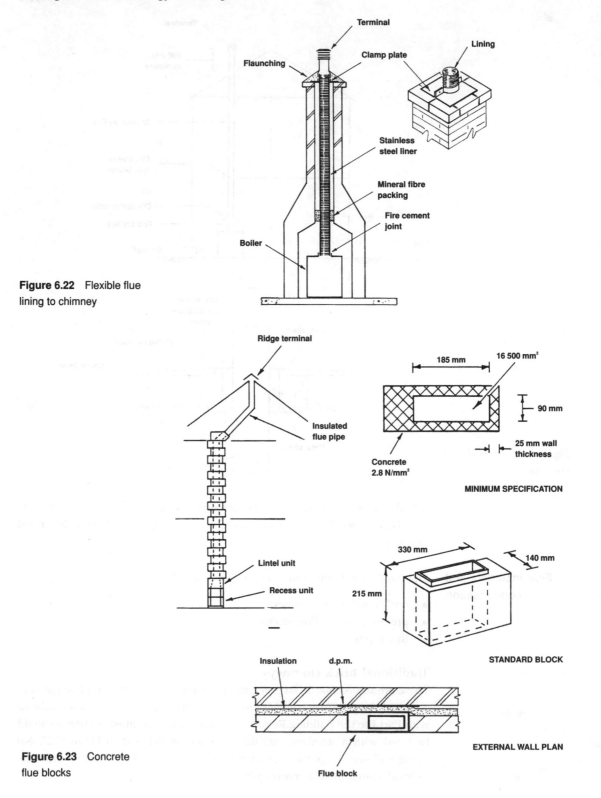

Figure 6.22 Flexible flue lining to chimney

Figure 6.23 Concrete flue blocks

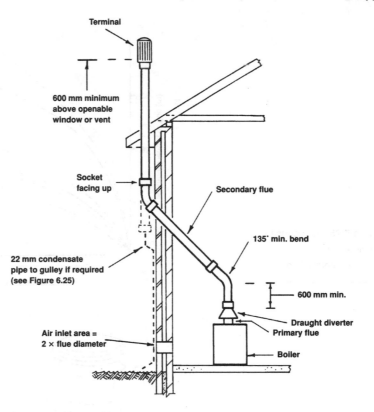

Figure 6.24
Conventional open flue

Precast concrete flue blocks

These are specially designed dense concrete flue blocks, with dimensions to bond in with the adjacent wall. They contain a void of minimum cross-sectional area 16 500 mm^2 to convey the flue gases from gas fires and convectors. Figure 6.23 details a standard block and the installation.

Flue pipes

Flue pipes are used internally or externally as shown in Figure 6.24. They are made in a range of sizes and materials to suit most situations, i.e. linings to existing chimneys or as free-standing flues. The most common materials are asbestos substitutes and enamelled or stainless steel. Double wall variations are preferred, having annular construction with the space between each pipe filled with insulation. These are particularly resistant to condensation and can extend to greater heights than single wall and masonry flues, as compared graphically in Figure 6.25.

Terminal location The flue terminal should promote the extraction of gases from the flue, resist the effects of down-draughts and prevent the ingress of rain and snow. The free area of outlet should be at least twice the flue area. The preferred location is at or above the ridge of a pitched roof, or at least 600 mm above the

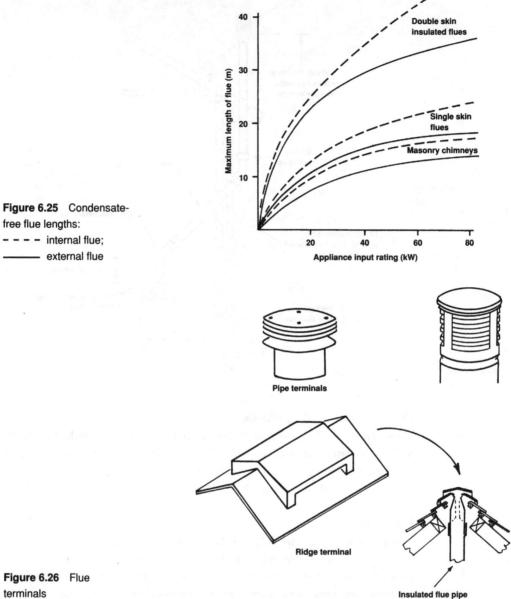

Figure 6.25 Condensate-free flue lengths:

− − − − internal flue;

─── external flue

Figure 6.26 Flue terminals

intersection in the pitch of a roof. Figure 6.26 shows typical ridge and pipe terminals, with Table 6.4 indicating heights of terminals in various situations.

Flues for large and high-rise buildings

Fan-diluted flue   In commercial and industrial premises where boilers are numerous and quite large, a correspondingly large flue extending from the plant room could be an

Table 6.4 Recommended open flue terminal location

Roof	Minimum height, roof line to terminal base
Flat, no parapets	250 mm
Flat, with parapets	600 mm*
Pitched >45°	1000 mm at or above ridge
Pitched <45°	600 mm at or above ridge

* Unless horizontal distance from parapet is more than 10 × parapet height, then 250 mm.

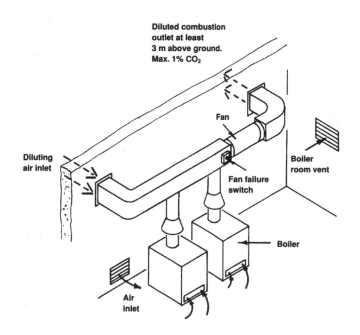

Figure 6.27 Fan-diluted flue to duplicate boilers

unsightly addition to the building. This can be overcome by fan dilution and horizontal discharge of the flue gases as shown in Figure 6.27. It is preferable for the diluting air inlet and diluted combustion gas outlet to be on the same wall to maintain a balance. The grille and terminal are well above headroom, at least 3 m from ground level. The relatively low height of discharge necessitates no more than a 1 per cent measured carbon dioxide content in the flue gases. Permanent ventilation to the boiler room is by louvred vents equivalent to twice the primary flue area. These are required to maintain efficient combustion and position at least 300 mm above ground level. In addition to flame failure devices and draught diverters, a fan safety device switches off the boilers if the fan fails to operate.

Shared flues   Shared flues are provided to simplify the disposal of burnt gas from numerous appliances in multi-storey buildings. Figures 6.28 and 6.29 show two

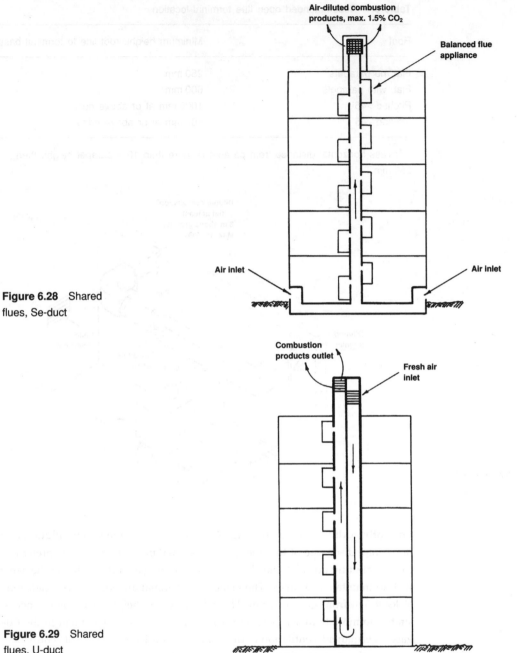

Figure 6.28 Shared flues, Se-duct

Figure 6.29 Shared flues, U-duct

applications, known as the Se-duct (originally developed by the SE Gas Board) and the U-duct respectively. Both are appropriate with room-sealed balanced flue appliances, the former drawing air for combustion at low level and the latter from the roof, requiring greater ducted area within the building. The vertical ducts are normally precast concrete sections, sized according to the

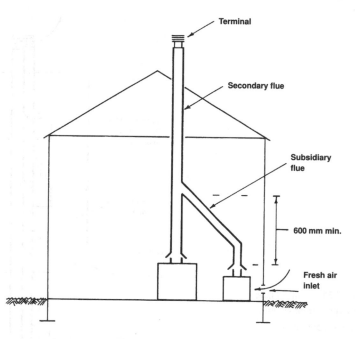

Terminal

Secondary flue

Subsidiary flue

600 mm min.

Fresh air inlet

Figure 6.30 Shared flues – appliances in same room.
Note: Consult gas authority when considering shared flues

number of appliances connected and design tables provided in BS 5440. Flame failure devices fitted to appliances prevent the build-up of unburnt gases in the ducts, and dilution of burnt gases measured at no more than 1.5 per cent carbon dioxide content at the uppermost appliance will ensure efficiency.

Branched flues   Several open or conventionally flued appliances (excepting gas fires) sharing the same room, can connect into a common flue as shown in Figure 6.30. The subsidiary flue must be at least 600 mm from the appliance draught diverter to its connection with the main flue.

Branched flues from appliances in different rooms or different parts of the same building are also acceptable, provided that all ventilation openings have the same aspect. Figure 6.31 shows the application, which with regard to varying wind pressures should not exceed 10 consecutive storeys. Water heaters and gas fires can be connected to this system provided the subsidiary flues are at least 1.2 and 3 m respectively, to ensure adequate draught.

Combustion analysis and fuel efficiency

Sophisticated instrumentation can be employed where precise details of carbon monoxide and carbon dioxide content in flue gases are required. For general applications, an adequate assessment can be made using the much simpler, portable Draeger analyser or the equally acceptable Fyrite analyser.

Draeger analyser   The equipment consists of neoprene hand bellows, expendable glass sampling tubes and a probe. The sampling tube is charged with crystals corresponding

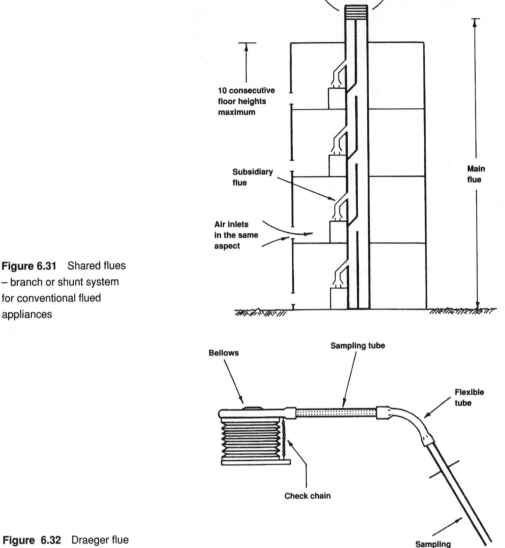

Figure 6.31 Shared flues – branch or shunt system for conventional flued appliances

Figure 6.32 Draeger flue gas analyser

to whether the carbon dioxide or carbon monoxide content is to be measured. With the probe inserted in the flue as shown in Figure 6.32, the bellows are pumped a number of times in accordance with the manufacturer's instructions. The crystals stain along a numerical scale, which converts on a chart to indicate percentage volume or parts per million.

Fyrite analyser This is primarily a carbon dioxide analyser and contains a reusable solution in a container, hand bellows, tube and sampling probe shown in Figure 6.33. The flue gas sample is pumped into the analyser, which is inverted to absorb

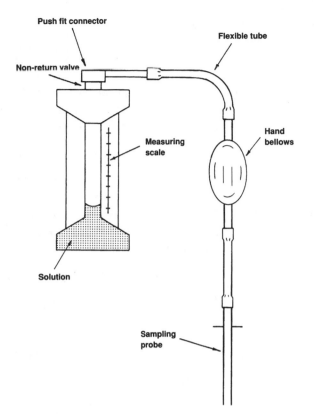

Push fit connector

Flexible tube

Non-return valve

Hand bellows

Measuring scale

Solution

Sampling probe

Figure 6.33 Fyrite flue gas analyser

the gas into solution. The liquid rises and the percentage carbon dioxide can be read from a scale on the side of the container. An alternative container and solution can be used to determine percentage oxygen if required. The Fyrite analyser is also useful for monitoring combustion efficiency of oil boilers.

Gas design calculations

Gas consumption Gas consumed by an appliance is established from the following formula:

$$Q = \frac{\text{Appliance input rating} \times 3600}{\text{Calorific value of the gas}}$$

where Q = quantity of gas in cubic metres per hour (m³/h). For example, a boiler of 18 kW input rating:

$$\text{Calorific value of natural gas} = 38\,000 \text{ kJ/m}^3$$

$$\text{Calorific value of LPG} = 95\,000 \text{ kJ/m}^3$$

For natural gas

$$Q = \frac{18 \times 3600}{38\,000} = 1.7 \text{ m}^3/\text{h}$$

For L.P.G.

$$Q = \frac{18 \times 3600}{95\,000} = 0.68 \, \text{m}^3/\text{h}$$

Operating costs Fuel tariffs are expressed in kilowatt hours (kWh). For the preceding example, the boiler operating from natural gas at 1.5 pence per kWh for 6 hours a day, the daily and weekly running costs will be

$$18 \, \text{kW} \times 6 \, \text{hours} \times 1.5\text{p} = £1.62$$

$$\text{For a 7 day week} = £11.34$$

Gas pressure Gas pressures are extremely low by comparison with other service systems and are measured in millibars or head in millimetres water gauge recorded on a manometer. The following provides a comparison:

$$1 \, \text{mbar} = 100 \, \text{N/m}^2 \text{ or pascals (Pa)} = 10 \, \text{mm w.g.} = 0.010 \, \text{m head}$$

For example, convert 3 kPa gas pressure to the more practical water gauge.

$$3 \, \text{kPa} = 3000 \, \text{Pa or } 3000 \, \text{N/m}^2$$

$$1 \, \text{mbar} = 100 \, \text{Pa or } 100 \, \text{N/m}^2$$

Therefore

$$\frac{3000}{100} = 30 \, \text{mbar}$$

$$\text{Head} = \frac{3000}{10^*} = 300 \, \text{mm water gauge}$$

* Actual figure is gravity of 9.81.

Pipe sizing The size of pipework depends on the gas consumption and the effective length of pipe. Tables 6.5 and 6.6 provide comparison for copper tube and steel pipes.

Table 6.5 Natural gas discharge rates through copper tube

Tube o.d. (mm)	Maximum effective length of tube (m)					
	3	6	9	12	15	20
6	0.12	0.06				
8	0.52	0.26	0.17	0.13	0.10	0.07
12	1.50	1.00	0.85	0.82	0.69	0.52
15	2.90	1.90	1.50	1.30	1.10	0.95
22	8.70	5.80	4.60	3.90	3.40	2.90
28	18.00	12.00	9.40	8.00	7.00	5.90

Table 6.6 Natural gas discharge rates through steel pipes

Pipe bore (mm)	Maximum effective length of pipe (m)					
	3	6	9	12	15	20
3	0.29	0.14	0.09	0.07	0.05	
6	0.80	0.53	0.49	0.36	0.29	0.22
9	2.10	1.40	1.10	0.93	0.81	0.70
12	4.30	2.90	2.30	2.00	1.70	1.50
20	9.70	6.60	5.30	4.50	3.90	3.30
25	18.00	12.00	10.00	8.5	7.5	6.3

Table 6.7 Resistance due to fittings

Pipe/tube diameter	Approximate equivalent lengths (m)		
	Elbow	Tee	90° bend
<25 mm bore steel and 28 mm o.d. copper	0.5	0.5	0.3

Example 1

Taking the previous example of an 18 kW boiler consuming 1.7 m³ of gas per hour, reference to Table 6.5 shows that a 15 mm outside diameter copper tube is adequate, provided the effective pipe length does not exceed 6 m.

Example 2

Figure 6.34 shows an installation to duplicate 35 kW input rated boilers from natural gas of calorific value 38 000 kJ/m³. Calculate the size of steel pipework, adding an allowance for resistance due to bends, etc. from Table 6.7.

Formula:

$$Q = \frac{\text{Appliance input rating} \times 3600}{\text{Calorific value of fuel}}$$

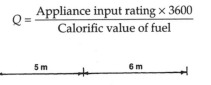

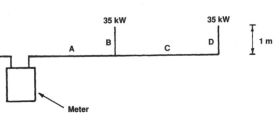

Figure 6.34 Gas pipe sizing

Pipe A:

$$Q = \frac{2 \times 35 \times 3600}{38\,000} = 6.63\,\text{m}^3/\text{h}$$

Effective pipe length <15 m, therefore pipe is 25 mm diameter.
Pipes B and D:

$$Q = \frac{35 \times 3600}{38\,000} = 3.31\,\text{m}^3/\text{h}$$

Effective pipe length <3 m, therefore pipe is 12 mm diameter.
Pipe C:

$$Q = \frac{35 \times 3600}{38\,000} = 3.31\,\text{m}^3/\text{h}$$

Effective pipe length <9 m, therefore pipe is 20 mm diameter.

7 Electricity supply and distribution

Background

During the early part of the twentieth century, electricity was produced and distributed by numerous small companies throughout the UK. Competition between these companies and the need for back-up generation plant, led to an irrational number of small power stations across the land, further complicated by variations in distribution and supply voltages.

Following the First World War, a central electricity board was formed to rationalise electricity generation and distribution by merging interests. This led to the creation of larger power stations, interconnected by a common grid. Thus, if a power station failed, there would be support and security of supplies from the others in the grid. The grid also:

- standardised electrical frequency and voltage for all consumers
- enabled transmission of very large voltages over long distances
- economised in plant and distribution
- permitted transfer of electricity throughout the country.

In 1948 the industry was nationalised in accordance with political trends of the time and duly became known as the Central Electricity Generating Board (CEGB) in 1957. It supplied electricity to 12 area authorities in England and Wales and had trading arrangements with the South of Scotland Electricity Board and the North of Scotland Hydro-Electric Board. Towards the end of the 1980s, the British government privatised the CEGB by splitting it into three companies, PowerGen, National Power and Nuclear Electric (now British Energy). The former two were responsible for oil-fired power stations, which with North Sea resources are gradually replacing coal-fired units, due to the closure of many pits. In Scotland the new generating companies are known as Scottish Power and the Scottish Hydro-Electric Power Company.

Supply

With the exception of hydroelectrical generation, fuel is converted to steam at high temperature and pressure, to drive turbines which generate electricity at 25 kV alternating current. The AC changes polarity from positive to negative at 50 cycles per second (50 hertz) and every generator in every power station is synchronised into the system at precisely the same frequency. The high

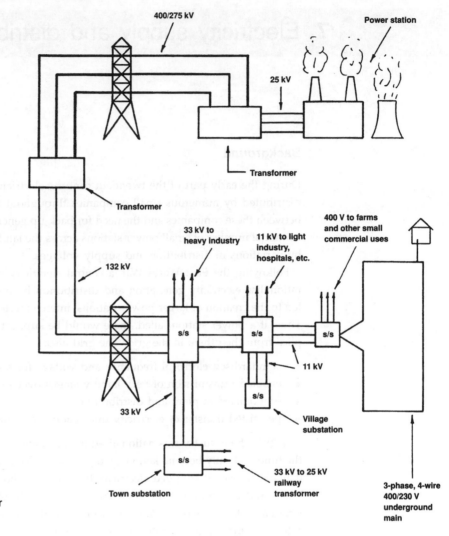

400/275 kV

Power station

25 kV

Transformer

Transformer

132 kV

**33 kV to
heavy industry**

**11 kV to light
industry,
hospitals, etc.**

**400 V to farms
and other small
commercial uses**

s/s

s/s

s/s

11 kV

33 kV

s/s

**Village
substation**

s/s

**33 kV to 25 kV
railway
transformer**

Town substation

**3-phase, 4-wire
400/230 V
underground
main**

Figure 7.1 Electricity
distribution.
Note: s/s = substation or
transformer

current produced enables the voltage to be transformed up to 132, 275 or
400 kV for national transmission. The former figure of 132 kV was the standard grid voltage of the 1930s through to the 1950s when it was uprated to
275 kV. In 1960 the 'supergrid' of 400 kV was conceived and has been introduced gradually to provide the carrying capacity needed to match demand
which doubles about every 10 years. Figure 7.1 shows how electricity is conveyed and transformed through substations from source to consumer. The
standard reduced voltages are:

400 000 ⎫
275 000 ⎬ Generation and distribution
132 000 ⎭
  33 000   Large/heavy industries, cities, towns and railways

11 000	Light industry, hospitals, towns and villages
400	Small industries, offices, farms, etc.
230	Housing, schools, small commercial premises

Note 1:   All but 230 V is supplied with three phases; domestic appliances and standard office equipment are designed to operate from only one phase.

Note 2:   33 000 V, three-phase supply is provided to railway operators via a substation, where it is transformed to 25 000 V direct current.

Note 3:   In addition to the grid, there is a 2000 MW direct current cross-Channel link with France to enable trade when peak demands vary. The positive and negative polarities neutralise the possibility of interference with ships' compasses.

Note 4:   Overhead lines or pylons may be unsightly, but are an economic necessity. The equivalent underground cable costs up to 20 times more to install.

Local distribution

Distribution and supply from 11 kV transformers or substations are shown in Figure 7.2. The incoming voltage is converted to 400 V potential in the three phases, which run with a neutral in one cable. The four-wire cable has phases plastic insulated in the colours red, yellow and blue for easy identification, with the neutral coloured black. These are normally buried about 1 m below the roads or pavings. Connections are shared across the three phases, so for an estate of 12 houses, 4 are on the red phase and 4 on each of the yellow and blue phases, with all connected to the neutral to provide 230 V potential. This is derived by dividing the three-phase voltage (400) by the square root of the (3) phases, i.e.

$$\frac{400}{1.73} = 231$$

Light industrial premises also receive a 230 V single-phase supply for lighting and small power applications. They may also require three-phase supply for machinery and this can be obtained by connecting across the remaining two phases to provide a potential of 400 V.

Intake – domestic   Most domestic supplies are buried about 0.5 m below ground and contain a phase and neutral in one cable, terminating at the meter cupboard shown in Figure 7.3.

In more remote areas, the supply may be overhead. Whatever the supply conditions, a provision must be made for connection with earth. The three most common supply systems are known as:

1.   TT (Figure 7.4)
2.   TN-S (Figure 7.5)
3.   TN-C-S (Figure 7.6).

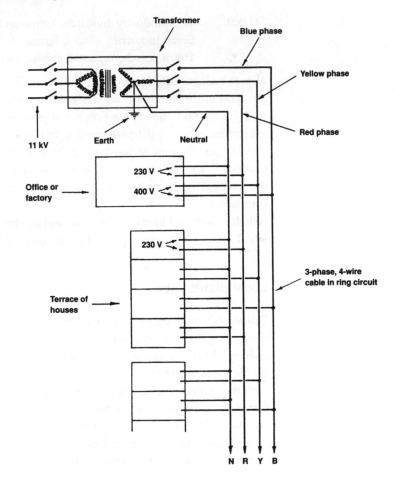

Figure 7.2 Three-phase and single-phase supply

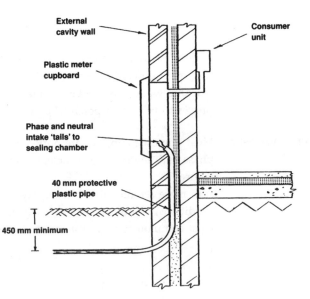

Figure 7.3 Service cable supply to external meter cupboard

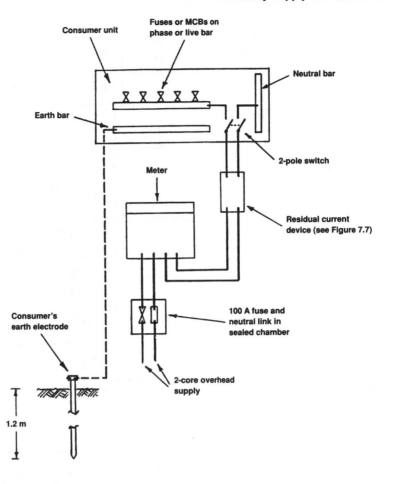

Figure 7.4 TT system

TT system

This is most likely to be used where the installation is supplied overhead, without an earth terminal. The consumer must provide an earth electrode, but as high resistance to earth may affect the connection, a residual current circuit breaker (RCCB, or trip switch) should be provided between the meter and the consumer unit. An RCCB operates on the principle of an installation fault causing an imbalance between the phase or live conductor and the neutral. As can be seen in Figure 7.4, a fault will energise the core coil to effect the electromagnetic trip.

TN-S system

This is applied to underground supplies which have metal sheathed and armoured cable. The consumer's earth terminal is connected to the metal sheath and this provides continuity back to the 11 kV transformer, where it is effectively connected to earth (see Figure 7.5).

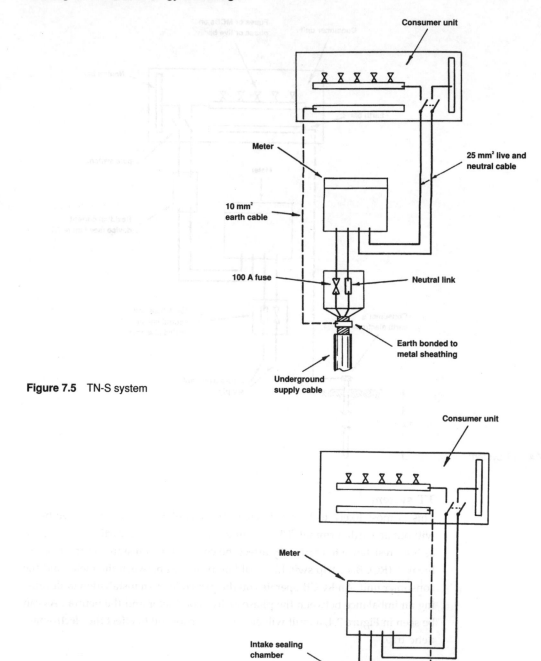

Figure 7.5 TN-S system

Figure 7.6 TN-C-S
system

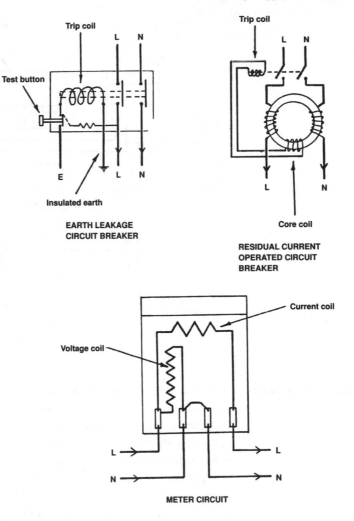

Figure 7.7 Circuit breakers and meter circuits

TN-C-S system

When the earth and neutral conductors combine to provide a protective earth and neutral (PEN), the system is known as TN-C, but if the consumer's installation has a separate neutral and earth connected to a TN-C supply, it is referred to as TN-C-S. Figure 7.6 shows the earth bar in the consumer unit connecting directly with the supply authority's neutral at the intake. Most new installations follow this mode, which is otherwise known as protective multiple earthing (PME).

Metering   Meter circuitry is shown in Figure 7.7. There are two coils, the current coil connected across the phase and the voltage coil across the phase and neutral. The interaction of the two energises a disc which rotates at speeds proportional to the power consumed. Gears driven by the disc effect a digital display of energy consumed in kWh.

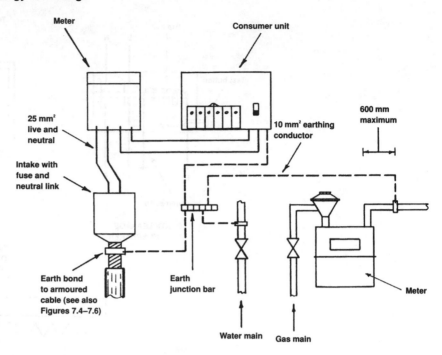

Figure 7.8 Bonding of services

Bonding of services It is essential that all metalwork associated with electrical appliances and installations, be cross-bonded to ensure the same potential as the electrical installation. This will prevent significant voltage differences at any point of contact. It should be undertaken in 10 mm² cable at the entry point of the services as shown in Figure 7.8.

Supplementary bonding may also be required where other extraneous metalwork exists. For instance, hot and cold water service pipes to bath, basin, sink, etc. are cross-bonded as shown in Figure 7.9 to protect a person making simultaneous contact with two electrically related pieces of equipment, when one may be faulty, e.g. an electric kettle and hot water tap.

Consumer unit The consumer's power supply control unit, conveniently summarised as a consumer unit, is a rationalisation of several circuit boxes containing a switch and fuse for isolation of individual circuits. The unit is located as close as possible to the meter, but on the inside of the building for convenient access. A two-pole main switch usually rated at 100 A controls the supply to several outgoing circuits or 'ways'. Each way is rated in amperes, the value depending on the circuit purpose. Circuit protection is by semi-enclosed rewirable fuse, cartridge fuse or miniature circuit breaker (MCB). Up to 16 ways are available for domestic use and a typical example is shown in Figure 7.10. Within the unit are a phase or live bar between fuses and isolator and an unfused neutral bar connected to the isolator. An unswitched earthing terminal is also provided.

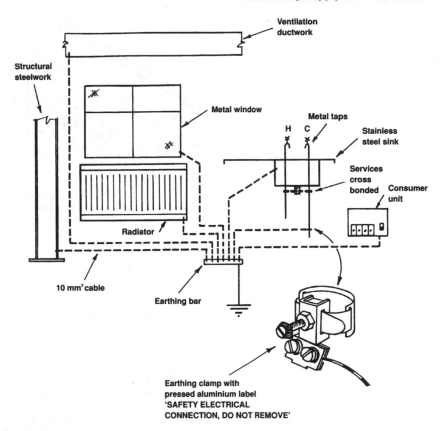

Figure 7.9 Examples of supplementary bonding

Variations exist for dual tariffs and split loads. The former are appropriate for cheap-rate overnight energy to storage radiators and the latter are used where an earth leakage circuit breaker (ELCB) provides specific overload protection to sockets on a ring main, while lighting is fed direct. Fuse and MCB ratings in a consumer unit vary. Traditionally 5, 15, 20, 30 and 45 A for fuses, but now 6, 10, 16, 20, 32, 40, 45 and 50 where MCBs are specified to European standards (BS EN 60898).

The general disposition in a consumer unit could be:

Lighting   6 A
Immersion heater   16 or 20 A (depending on rating)
Power socket ring main   32 A
Cooker   40 or 45 A (depending on rating)
Shower   40 or 45 A (depending on rating)

Overload protection This can be provided by:

- semi-enclosed rewirable fuses (Figure 7.11)
- cartridge fuses (Figure 7.12), or
- miniature circuit breakers (Figure 7.13).

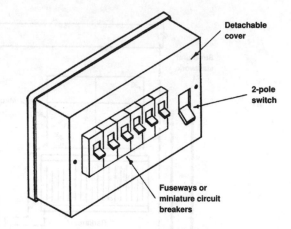

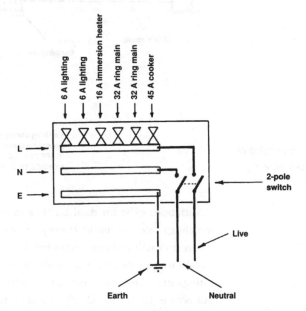

Figure 7.10 Standard six-way consumer unit

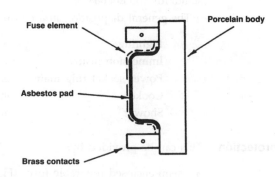

Figure 7.11 Rewirable fuse

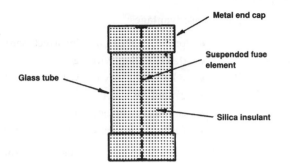

Figure 7.12 Cartridge fuse

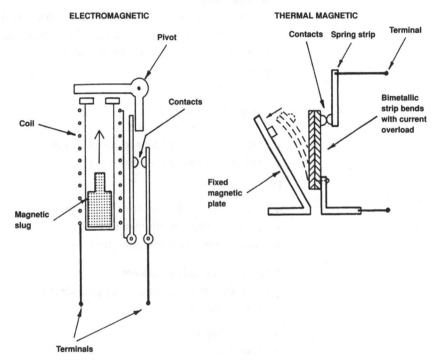

Figure 7.13 Operation of miniature circuit breakers

Semi-enclosed rewirable fuse

This consists of a plastic or porcelain base and carrier for the wire. The base is screwed into the consumer unit and has fixed contacts to the live bar and circuit cable. Sometimes the fuse wire has an asbestos tape base or tube, to prevent damage by arcing when the fuse ruptures.

Advantages:

- low capital cost
- cheap to replace the element
- simple in concept
- no moving parts.

Disadvantages:

- prone to abuse, i.e. incorrect wire can be used
- deterioration with age
- reliability can vary, with temperature and method of fixing
- cannot be tested
- cannot be replaced quickly
- fusing factor may be much higher than current rating, i.e.

$$\frac{\text{Fusing current}}{\text{Fuse current rating}} = \text{Fusing factor}$$

Cartridge fuses

These consist of two metal end caps, to which the fuse element is attached. The element is enclosed in a glass tube containing silica insulant. Different diameters correspond to different ratings, so that incorrect substitution is impossible.

Advantages:

- small and compact
- no moving parts
- accurate and consistent to declared rating
- will not deteriorate in protective tube.

Disadvantages:

- cannot be repaired or reset, must be replaced
- prone to abuse (with metal foil!)
- much more expensive than fuse wire.

Miniature circuit breakers

These small automatic circuit tripping devices are gradually replacing fuses. Their operation is either:

- electromagnetic, or
- thermal.

Magnetic tripping units have a current passing through an electromagnet which in normal use is insufficient to cause a magnetic pull. A surge of current due to an overload energises the magnet to break contact physically or by magnetic field.

Thermal tripping is effected with a bimetallic strip. Current overload causes the strip to warm and as it moves, a fixed magnetic plate or electromagnet accelerates the reaction.

Cables

PVC sheathed and insulated
PVC sheathed and insulated cable has replaced the very dated lead sheathed and rubber insulated cable, occasionally found in unmodernised older buildings.

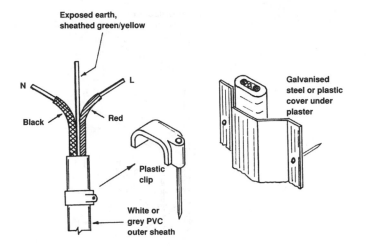

Figure 7.14 PVC
sheathed cable

Rubber sheathed cable may also be found, but this too has been superseded by PVC as rubber is less resilient to chemicals, oil and direct sunlight.

PVC insulated cables are manufactured with one, two or three conductors, with or without an earth wire. Earth conductors are normally bare and uninsulated, but where exposed in junction boxes, light switches, etc. must be sheathed with green and yellow striped PVC tubing for identification. The insulation also prevents unnecessarily earthing a circuit if they were to contact live or neutral terminals. Live or phase conductors are insulated in red PVC and neutral, black. Blue and yellow insulated live conductors are occasionally used in addition to red, in instances such as two-way lighting. Figure 7.14 shows a standard white outer sheathed PVC cable with live (red), neutral (black) and earth (exposed) conductors with clips or a protective cover for use under plaster. Steel or plastic tube conduit and trunking shown in Figure 7.15 may also be used to provide continuity of support and protection from physical damage. Some protection from thermal damage will also be provided by conduit, but in high-temperature situations, mineral insulated copper covered cables (MICC) should be specified.

Mineral insulated copper covered

These cables have copper conductors located within highly compressed powdered magnesium oxide. This acts as an insulant and with the outer copper protective sheath, it can resist temperatures up to 150 °C. Cutting and termination require special procedures to prevent atmospheric dampness penetrating the insulation. After cutting, the magnesium oxide is immediately sealed with a compound and fitted with a neoprene sleeve and anchor as shown in Figure 7.16.

Cable rating

Cable specification must be sufficiently generous to limit the heating effect caused by the electrons flowing through the conductor. An underrated cable will offer excessive resistance and the energy generated could damage the

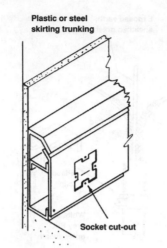

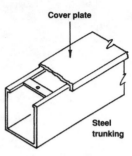

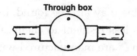

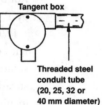

Figure 7.15 Trunking and conduit

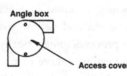

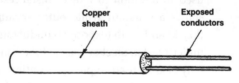

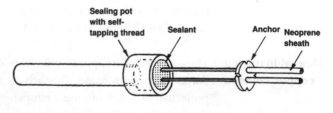

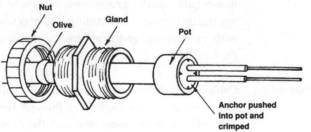

Figure 7.16 Termination of mineral insulated cable (MICC)

Table 7.1 Cable rating and standard applications

Application	Cable cross-section (mm²)	Minimum overload protection (A)	Comments and limitations
Lighting	1.0 or 1.5	5	Max. 10 lights
Immersion heater	1.5 or 2.5	15	Heat-resisting flex from fused socket to element
Sockets on radial circuit	2.5	20	Max. 20 m² floor area and 17 m of cable
	4.0	30	Max. 50 m² floor area and 21 m of cable
Sockets on ring main	2.5	30	Max. 100 m² floor area and 54 m of cable
Cooker	6.0 or 10	30 or 45	Cable and fuse specification depends on cooker rating
Shower	4 (7.2 kW)	45	4 mm² cable, max. 18 m length
	6 (8.3 kW)	45	6 mm² cable, max. 27 m length (TN-C-S system)

cable insulation. The following factors should be considered before selecting cable for a specific situation:

- the appliance load on the circuit and maximum current expected
- the length of cable and associated voltage drop
- location of cable, whether exposed or protected
- whether the cable is closely grouped with others
- air temperature surrounding the cable
- the overload protection provided.

For small circuits and domestic installations using PVC sheathed and insulated cables, the specification is fairly standard as listed in Table 7.1.

Cable sizing Where unusual or special installations are needed, the cable specification must be calculated. This can be established by determining the design current of a circuit and relating this to the cable current capacities shown in Table 7.2. A check that the fitted appliances will not suffer a voltage drop in excess of 4 per cent is also necessary. The procedure is as follows.

Table 7.2 Properties of PVC insulated copper conductor cables

Conductor cross-sectional area (mm²)	Current carrying capacity (A)		Voltage drop per amp per metre (mV)
	Loose clipped	In conduit	
1.00	15.5	13.5	44.0
1.50	20.0	17.5	29.0
2.50	27.0	24.0	18.0
4.00	37.0	32.0	11.0
6.00	47.0	41.0	7.3
10.00	65.0	57.0	4.4
16.00	87.0	76.0	2.8
25.00	114.0	101.0	1.8
35.00	141.0	125.0	1.3

Note: Maximum voltage drop allowed is 4 per cent of the supply voltage.

Example 1

A 6 kW motor is to operate from a 230 V single-phase cable, 10 m long.

1. Find the current flowing:

$$\text{Amps} = \frac{\text{Watts}}{\text{Volts}} = \frac{6000}{230} = 26 \text{ A}$$

2. From Table 7.2 select a suitable cable, i.e. 2.5 mm² loose clipped (27 A) or 4 mm² in conduit (32 A).

3. Check voltage drop (4 per cent max.):

$$\text{Volt drop} = \frac{\text{mV} \times \text{current flowing} \times \text{cable length}}{1000}$$

$$2.5 \text{ mm}^2 \text{ (loose)} = \frac{18 \times 26 \times 10}{1000} = 4.68 \text{ V}$$

$$4.0 \text{ mm}^2 \text{ (in conduit)} = \frac{11 \times 26 \times 10}{1000} = 2.86 \text{ V}$$

4. Maximum voltage drop acceptable is

$$230 \times \frac{4}{100} = 9.2 \text{ V}$$

therefore both cables are satisfactory.

Example 2

A 3 kW load from a 230 V single-phase loose clipped supply cable, 30 m long.

1.
$$\text{Amps} = \frac{\text{Watts}}{\text{Volts}} = \frac{3000}{230} = 13 \text{ A}$$

2. From Table 7.2, select a suitable cable, i.e. $1.00 \, \text{mm}^2$ (15.5 A) appears satisfactory.
3. Check voltage drop (4 per cent max.):

$$\text{Volt drop} = \frac{44 \times 13 \times 30}{1000} = 17.16 \, \text{V}$$

4. Maximum voltage drop acceptable = 9.2 V (see Example 1), therefore $1.00 \, \text{m}^2$ is unacceptable.
5. Try a higher specification cable: $1.5 \, \text{mm}^2$ is also unacceptable, with a voltage drop of 11.3 V, but $2.5 \, \text{mm}^2$ is acceptable, i.e.

$$\text{Volt drop} = \frac{18 \times 13 \times 30}{1000} = 7 \, \text{V}$$

Lighting circuits

Small subcircuits of the type used in housing are limited to a load of 1 kW, i.e. 10 light fittings of standard 100 W energy rating. Using the formula

$$\text{Amps} = \frac{\text{Watts}}{\text{Volts}} = \frac{1000}{230} = 4.3$$

it can be seen that 5 or 6 A overload protection in the consumer unit will be adequate for each circuit. From Table 7.2, a minimum cable specification of $1.00 \, \text{mm}^2$ is sufficient, subject to checking voltage drop. Most dwellings will have two lighting circuits, one for each floor. In larger buildings, where a greater number of light fittings are required, fuse ratings of up to 16 A are used, with correspondingly higher-rated cable.

There are three methods adopted for wiring lights:

1. The junction/joint box system
2. The loop-in system
3. The combination of (1) and (2).

Junction or joint boxes This arrangement is less economic than the loop-in system and is found mainly in older installations. Figure 7.17 shows the use of 4 terminal joint boxes, 3 terminal ceiling roses with flex to bulb holders and one-way switching continuing up to the maximum of 10 fittings.

Loop-in This is the most common of contemporary installations, being cheaper and simpler than using joint boxes. There are less connections to make, which is more convenient, quicker and safer for both sheathed cable and conduit installations. Figure 7.18 shows a typical system using four terminal ceiling roses and one-way switching.

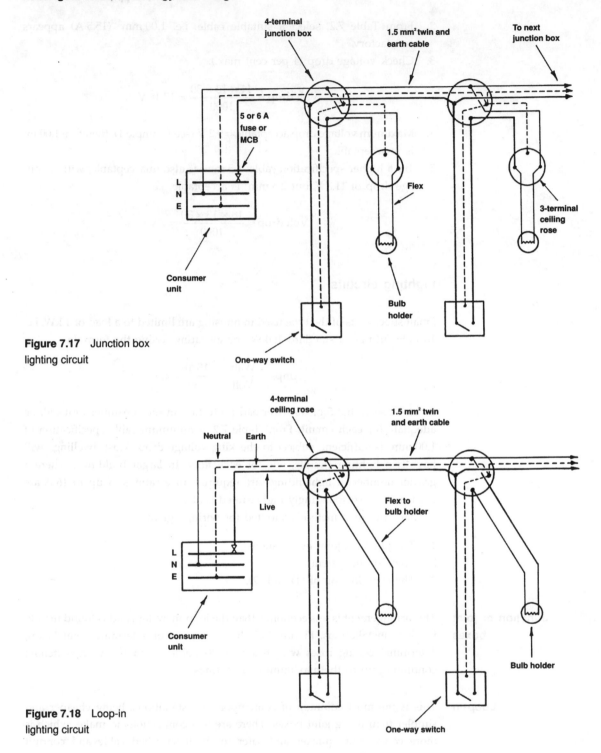

Figure 7.17 Junction box lighting circuit

Figure 7.18 Loop-in lighting circuit

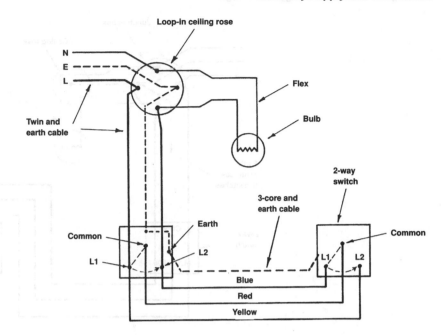

Figure 7.19 Two-way switching

Combination of joint boxes and loop-in

This is unlikely to be planned for an initial installation, but a loop-in system may be extended to a joint box to facilitate some extra fittings. Alternatively, some lighting points in a predominantly loop-in system may be connected through joint boxes to save cable.

Flex to bulb holders

Heat-resisting flex, insulated with butyl, ethylene propylene rubber or silicone rubber is used between ceiling rose and bulb holder. PVC insulated cable is unsuited to this situation as it will become brittle and crack in the presence of heat from light bulbs. The colour coding of flex should not be confused with that of circuit cable. Brown corresponds to live, blue to neutral and green with yellow stripe is earth.

Note: Switches for bathroom illumination should be outside the room or if inside, operated by remote string pull to prevent direct contact with damp fingers.

Two-way and intermediate switching

Multiple controls for one or more light fittings are convenient in hall and landing situations and bedrooms, with light switch near the door and string pull switch over the bed. Special two-way and intermediate switches are installed with three-core and earth cable in-between. Figure 7.19 shows wiring for a typical domestic hall and landing from the last ceiling rose in a loop-in system and Figure 7.20 shows an intermediate system suited to long corridors and multi-flight stairs. Additional intermediate switching can be provided if required.

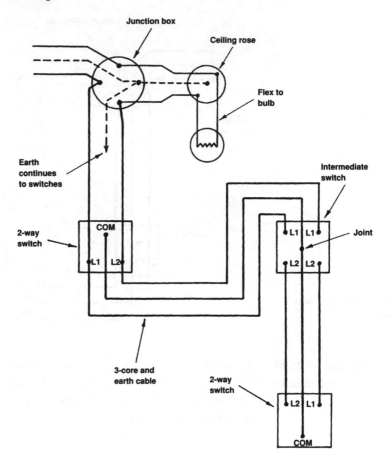

Figure 7.20 Intermediate switching

Power sockets

Sockets should be provided generously and preferably of the double pattern type to reduce the use of adaptors. Quantity and location in commercial and industrial premises will depend on the building's function, but for domestic purposes, Table 7.3 provides guidance for various rooms.

Sockets are placed between 150 and 250 mm above work surfaces and at least 150 mm above floor level. In buildings especially designed for the elderly and disabled, a socket height of between 750 and 900 mm is recommended. Disposition should be such that flexible cord to portable appliances will not exceed 2 m. Flexes connect to sockets by means of a three-pin plug containing a small cartridge fuse. Most modern appliances are pre-wired to the plug, but where it is necessary to attach the plug, the brown insulated cable connects to the live terminal, the blue to the neutral and green with yellow to the earth. Fuses inserted on the live terminal are selected with regard to the appliance load, as listed in Table 7.4.

Table 7.3 Recommended provision for power sockets in housing

Location	Number
Kitchen	6
Utility room	3
Living rooms	8
Dining room	4
Master bedroom	6
Single bedrooms	3
Study bedroom	4
Hall and landing	2
Garage/workshop	1
Bathroom	1 shaver socket (double insulated)

Table 7.4 Equivalent current and power ratings at 230 V

Fuse rating (A)	Appliance load (W)
1	230
2	460
3	690
5	1150
7	1610
10	2300
13	2990

Derived from:

$$\text{Amperes} = \frac{\text{Wattage}}{\text{Voltage}}$$

Standard circuits

The final circuit to power sockets may be:

- radial, or
- ring main.

Radial

The radial circuit shown in Figure 7.21 can have an unlimited number of sockets, provided the overload protection at the consumer unit, the floor area served and cable length are in accordance with the details in Table 7.1. Although sockets are unlimited, the estimated maximum demand on the circuit must not exceed the rating of the overload protection device (fuse or MCB) in the consumer unit.

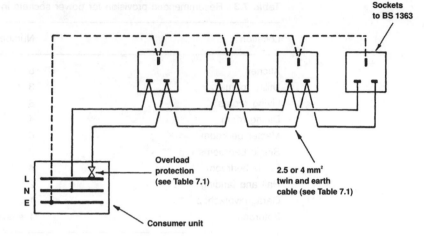

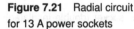

Figure 7.21 Radial circuit for 13 A power sockets

Ring

A ring circuit, as the name suggests, has one cable containing live or phase, neutral and earth completing a ring, commencing and finishing at the consumer unit. Both phase conductors connect to a 30 or 32 A fuse or MCB, with neutrals and earths connecting to respective terminals. Although the cable rating (see Table 7.2) is less than the fuse rating, it will not be overloaded since the current splits both ways around the ring to provide even distribution to the sockets. An unlimited number of sockets can be installed within the constraints listed in Table 7.1 and provided the estimated maximum circuit demand does not exceed the overload protection. Hence the relatively heavy machinery used in domestic kitchens will warrant a dedicated ring main for this area, plus a further ring main for each floor level.

Figure 7.22 shows the principle of a ring main, with some additional spur sockets. These sockets are limited and should not exceed, in number, the total connected direct to the ring. They may connect at the terminals of a ring socket or from a joint box attached to the ring. A fused spur must be used if the end appliance is fixed, e.g. an inset fire.

Fixed appliances

Many fixed appliances such as clocks, fans, fires, central heating controls and small water heaters can be connected to a fused spur on a ring main. Appliances and loads generally rated over 3 kW are provided with separate radial circuits from the consumer unit. These include:

1. Cooker
2. Water heaters:
    (a) immersion
    (b) instantaneous

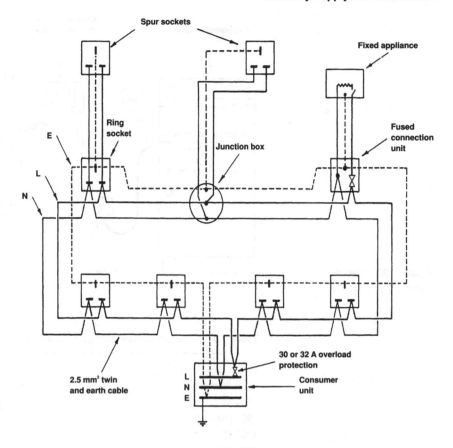

Figure 7.22 Ring circuit for 13 A power sockets

3. Off-peak:
   (a) underfloor heating
   (b) storage radiators.

Cooker Cooker ratings vary, but a total load of about 12 kW is normal for a domestic appliance and may be exceeded with a separate hob and oven. Smaller free-standing units are adequately supplied from a 30 or 32 A fuse/MCB with 4.0 mm² cable. Most modern units require a 45 A overload protection through 6.0 mm² or even 10.0 mm² cable, terminating at a 50 A rated double pole control switch. A simple calculation suggests that 30 A or even 45 A overload protection is inadequate, as a 12 kW cooker plus 3 kW kettle plugged into the cooker socket could demand over 65 A! In reality, not all plates, ovens, fan, etc. are likely to be on at any one time. Furthermore, thermostats will regulate oven output, so actual current demand will be considerably lower than 65 A and comfortably within the preceding specified data. This logic is known as diversity and forms the basis for design and assessment of all electrical installation work.

The cable run should be as short and direct as possible, terminating at a control unit not less than 1.35 m above floor level and slightly to one side of the cooker as shown in Figure 7.23.

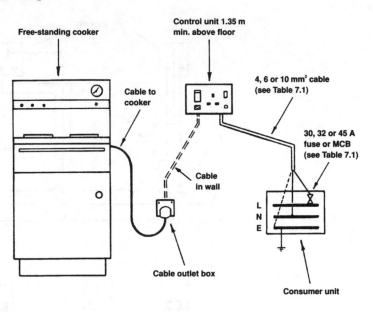

Figure 7.23 Radial supply to cooker

Table 7.5 Immersion heater ratings for various water vessel capacities

Water capacity (litres)	Rating (kW)
70–90	1
90–110	2
110–135	3
135–220	4

Water heaters

Immersion unit

Immersion heaters are an insulated electric element, classified as a continuous load. They usually rate at 3 kW and therefore would impose too much if they spurred off the ring main. Table 7.5 lists other ratings and corresponding hot water storage vessel capacities.

Cable rated at 1.5 mm² is sufficient, but 2.5 mm² is preferred for 15 or 16 A overload protection. A 4 kW element will require 20 A protection. An essential control is a double pole isolator switch close to the unit, preferably fitted with a neon indicator. Heat-resisting butyl rubber coated flex connects switch to element terminals via a thermostat set at about 65–70 °C. Figure 7.24 shows the installation to a typical domestic hot water cylinder, with details of connections to element and thermostat.

Some immersion heaters are double element, with a short 500 W unit to heat the upper part of the cylinder only. This is adequate for sink and basin use and when a bath is required, the more powerful element can be switched on. An alternative uses two separate heating elements as shown in Figure 7.25.

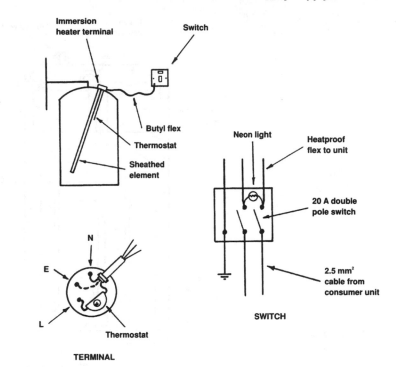

Figure 7.24 Radial
supply to immersion heater

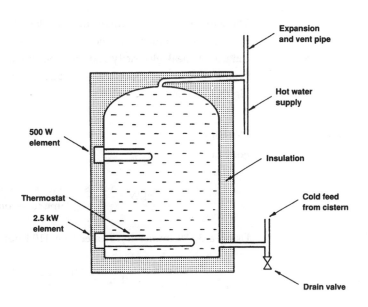

Figure 7.25 Double
element water heater

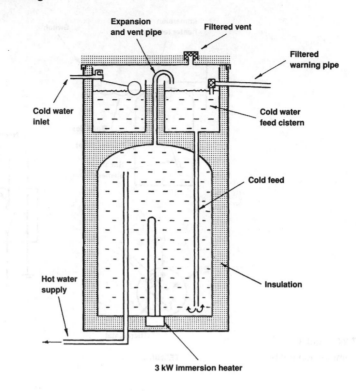

Expansion and vent pipe

Filtered vent

Filtered warning pipe

Cold water inlet

Cold water feed cistern

Cold feed

Hot water supply

Insulation

3 kW immersion heater

Figure 7.26 Combined cistern and hot water storage unit

These are a resourceful application in flats and being relatively small, can be discreetly accommodated under a sink unit or other cupboard location. They can use off-peak electricity and will complement electric underfloor or storage heaters. If there is space for a storage cistern, the combined unit shown in Figure 7.26 could be used.

Costing
Formula:

$$kW = \frac{\text{Litres} \times \text{s.h.c.} \times \text{temperature rise}}{\text{Time in seconds}}$$

Using a 3 kW immersion heater in a 110 l hot water storage cylinder,

$$\text{Time} = \frac{110 \times 4.2 \times 40}{3} \qquad = 6160 \text{ s}$$

$$= 1.7 \text{ hours} \times 3 \text{ kW} \quad = 5.1 \text{ kWh}$$

$$5.1 \text{ kWh at 7p/unit} = 36 \text{ pence}$$

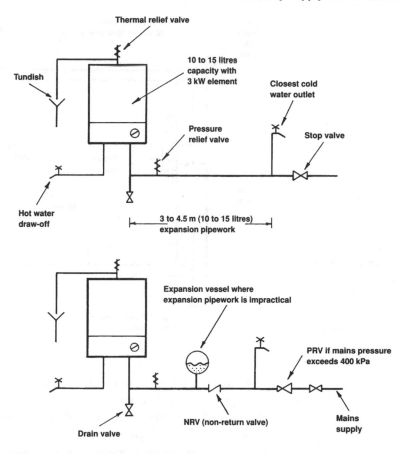

Figure 7.27 Unvented water heaters

Instantaneous water heaters

These are mains-fed units fitted with a safety pressure switch to prevent the element switching on and overheating in the absence of water. They also have a thermal cut-out to prevent the water overheating and scalding the user.

Basins and sinks

Figure 7.27 shows a simple 3 kW unit applied over a basin or sink. It can be supplied from a fused spur on the ring main or from an independent radial circuit. The mains water cold feed must incorporate a pressure relief valve (PRV) and at least 3 m of 15 mm copper tube (or equivalent) to absorb hot water expansion in the pipework. If this is impractical, an expansion vessel can be used for the same purpose and must be used where the inlet pressure is over 4 bar (400 kPa).

Showers

These are rated between 5 and 9.5 kW and therefore require a fuse or MCB rating of between 30 and 45 A at the consumer unit. The radial supply cable may

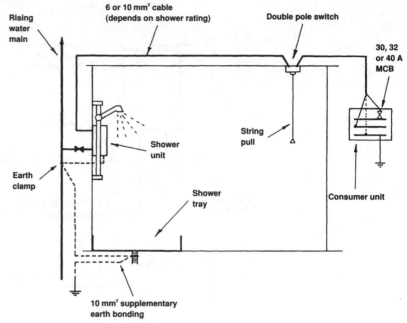

Rising
water
main

6 or 10 mm² cable
(depends on shower rating)

Double pole switch

30, 32
or 40 A
MCB

String
pull

Shower
unit

Earth
clamp

Shower
tray

Consumer unit

10 mm² supplementary
earth bonding

Figure 7.28
Instantaneous shower.
MCB:
30 A for 7 kW element;
32 A for 7.5 kW element;
40 A for 8–9.5 kW element

be 4.0 mm² for the lower loadings, but 6.00 mm² should be used in the event of a higher-rated replacement unit being fitted at a later date. The highest-rated showers will need 10 mm² and manufacturers' literature should be consulted to ensure the correct specification. Isolation is by readily accessible double pole switch, of the cord-operated type. If cord operation is impractical, a standard two-pole switch can be provided, but out of reach of the person using the shower. Figure 7.28 shows the preferred means of switching, and details of electrical and mains water supply.

**Off-peak heating
(Economy 7)**

Electricity generators' plant runs for 24 hours a day, with considerable spare capacity overnight. To encourage the use of this surplus, it is sold at a cheaper rate than normal and timers can be incorporated at the intake to switch to a cheap-rate meter, somewhere between 2300 and 0700 hours. Dishwashers, washing machines, etc. can be timed to take advantage of the cheaper rate, but the greatest benefit is to absorb the energy as thermal capacity within the floor structure as underfloor heating, or within storage radiators for dissipation throughout the next day.

Underfloor heating

This system uses high-resisting insulated conductors embedded in a screed approximately 50 mm below the surface. Conductors are spaced between 100

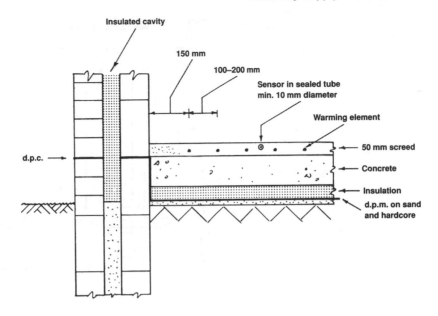

Figure 7.29 Underfloor heating

and 200 mm depending on required output, which ranges between 10 and 20 W/m, depending on cable specification. To succeed, the underside of the floor structure needs insulation, now complemented by mandatory provision in the Building Regulations. Figure 7.29 shows the installation within a screed and Figure 7.30 shows the control concept and floor distribution using both twin and single conductors.

Night storage radiators

These are free-standing cabinets containing electric elements inserted between refractory concrete blocks. Originally they were criticised for their bulk, but now with improvements in storage block material and with fan-assisted delivery, they occupy little more space than conventional hot water radiators. An input control thermostat on each radiator is set manually, with regard to outside temperatures, and an internal thermostat prevents an excess of heat building up. Fan delivery is usually on a time clock and can be switched off as required, to conserve some of the heat within the blocks. Units are rated between about 1 and 6 kW and can be selected from manufacturers' design tables. These incorporate room details such as floor area, room height, window area, etc. As a rough guide, 200 W output is required for every square metre of floor in a typical modern estate house. For example, for a living room measuring 5 m × 5 m, the load will be 5 m × 5 m × 200 W = 5 kW. One 5 kW unit or for better heat distribution, two at 2.5 kW.

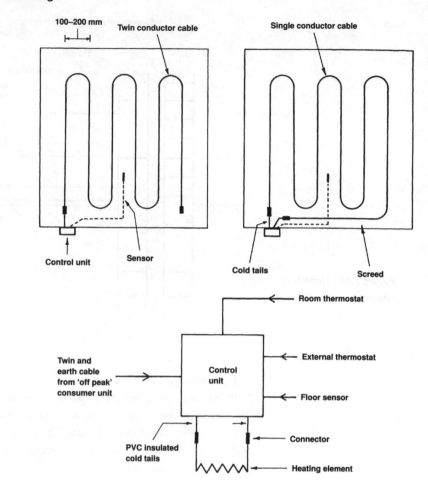

100–200 mm

Twin conductor cable

Single conductor cable

Control unit   Sensor

Cold tails   Screed

Room thermostat

Twin and
earth cable
from 'off peak'
consumer unit

Control
unit

External thermostat

Floor sensor

PVC insulated
cold tails

Connector

Heating element

Figure 7.30 Installation of underfloor heating elements

Figure 7.31 shows a separate radial circuit for each storage radiator. As all storage radiators switch on together, they cannot be supplied from a ring circuit as this relies on diversity (variable demand) of the connected appliances. A 2.5 mm^2 cable is used, with a 20 A fuse or MCB at the time-controlled consumer unit. A 20 A double pole control switch with heat-resisting flex outlet is provided adjacent to each heater.

External extensions

Supply to a shed, garage or greenhouse detached from the main building can be provided overhead or underground. Ordinary PVC sheathed cable can be used and the minimum recommended height is 3.5 m. If the span exceeds 3 m,

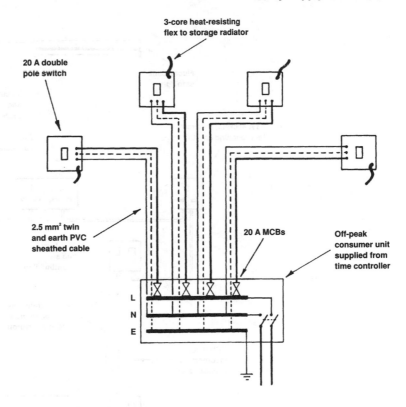

3-core heat-resisting
flex to storage radiator

20 A double
pole switch

2.5 mm² twin
and earth PVC
sheathed cable

20 A MCBs

Off-peak
consumer unit
supplied from
time controller

L

N

E

Figure 7.31 Radial
circuits to four storage
radiators

support must be provided by continuous conduit or a catenary wire suspended from both buildings. Underground is unobtrusive, but should be at least 500 mm below the surface in armoured PVC insulated cable or copper sheathed MICC. Ordinary PVC insulated cable is satisfactory if enclosed in galvanised steel or plastic conduit. A switch/fuse unit is located close to a spare fuseway in the main consumer unit. The overload protection and cable rating will relate to the anticipated end use, but for nominal power and light, a 20 A MCB will be sufficient with 2.5 mm² cable. However, if 30 A protection is necessary the cable must be 4.0 mm² and can serve up to five sockets and a 3 A light circuit as shown in Figure 7.32.

Distribution design and diversity factors

If cables and overload protection devices were rated for the full capacity of every subcircuit, size of components and costs would be impractical, prohibitive and unnecessary. Therefore the Institute of Electrical Engineers' (IEE) Regulations permit diversity factors or allowances for partial use of some facilities and equipment to be incorporated into the calculations, when assessing maximum demand.

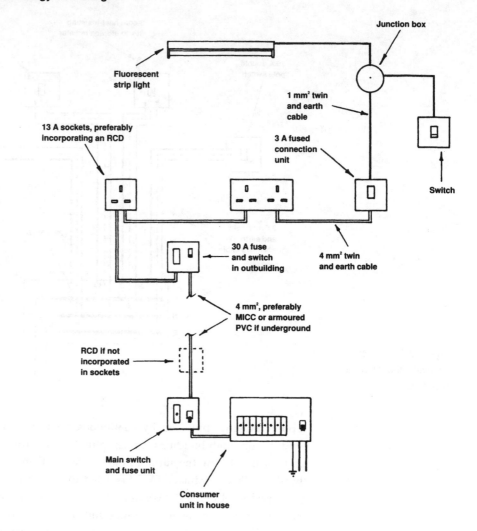

Figure 7.32 Radial supply to outbuilding

Example 1 A typical three-bedroomed estate house, with single phase 230 V supply, having:

1. Two lighting circuits of 900 W each
2. Three 30 A ring mains to 13 A power sockets
3. One 3 kW immersion heater
4. One 12 kW cooker with one power socket.

Application	Current demand (A)	Diversity factor*	Diversified current demand (A)
1. 2×900 W			
$= 1800$ W			
$\dfrac{1800}{230} =$ 7.83		66%	5.16
2. Circuit 1	30.00	100%	30.00
Circuit 2	30.00	40%	12.00
Circuit 3	30.00	40%	12.00
3. $\dfrac{3000}{230} =$ 13.04		100%	13.04
4. $\dfrac{12\,000}{230} =$ 52.17		10 A + 30% of remainder (42.17), + 5 A for socket	10.00 12.65 5.00
		Total =	99.85 A

* Consult IEE Regulations.

Example 2   A 12-bedroomed hotel with 230 V, 3-phase supply having:

1. Four lighting circuits of 800 W each
2. Four 30 A ring mains
3. Two 3 kW immersion heaters
4. Three 12 kW cookers.

Application	Current demand (A)	Diversity factor*	Diversified current demand (A)
1. 4×800 W			
$= 3200$ W			
$\dfrac{3200}{230} =$ 13.91		75%	10.43
2. Circuit 1	30.00	100%	30.00
Circuit 2	30.00	50%	15.00
Circuit 3	30.00	50%	15.00
Circuit 4	30.00	50%	15.00
3. $\dfrac{6000}{230} =$ 26.08		100%	26.08

Application	Current demand (A)	Diversity factor*	Diversified current demand (A)
4. $\dfrac{12\,000}{230} =$	52.17	100%	52.17
$\dfrac{12\,000}{230} =$	52.17	80%	41.74
$\dfrac{12\,000}{230} =$	52.17	60%	31.30
		Total =	236.72 A

$$\text{Spread over 3 phases} = \frac{236.72}{3} = 79\ \text{A per phase}$$

* Diversity factors and allowances vary for different applications and building types. Full details can be found in the IEE Regulations.

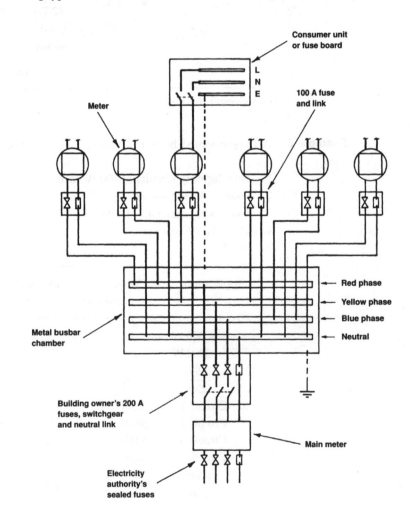

Figure 7.33 Three-phase radial distribution to low-rise flats or offices

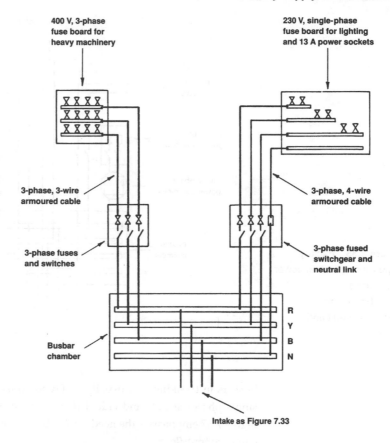

400 V, 3-phase
fuse board for
heavy machinery

230 V, single-phase
fuse board for lighting
and 13 A power sockets

3-phase, 3-wire
armoured cable

3-phase, 4-wire
armoured cable

3-phase fuses
and switches

3-phase fused
switchgear and
neutral link

R
Y
B
N

Busbar
chamber

Intake as Figure 7.33

Figure 7.34 Three-phase radial distribution to factory premises.
Note: Earth continuity through metal conduit and trunking

Supply to large and tall buildings

The electrical demand in large buildings such as blocks of flats, factories, offices, etc. will necessitate an intake of all three phases and a neutral to operate and satisfy a wide distribution of electricity to machinery, lifts and multiple banks of lighting. The largest of buildings and campuses will have their own 11 kV substation and transformers to variously disposed subsidiary systems.

Large buildings The intake and distribution in large buildings can be of radial pattern shown in Figures 7.33 and 7.34. Here each of the three phases and neutral have fused switchgear at the intake and connections to a horizontal busbar chamber. In factories the chamber or trunking runs around the building interior for convenient access for cables to be bolted to the exposed copper busbars. The other advantage of a busbar chamber is division of the phases, which may be used to supply separate connections and metering for various facilities. This is shown in Figure 7.35 with supply to fuse boards for heating, lighting, power, etc.

Tall buildings In tall buildings an insulated three phase and neutral can extend the full height, with connections taken at each floor. However, for convenient access,

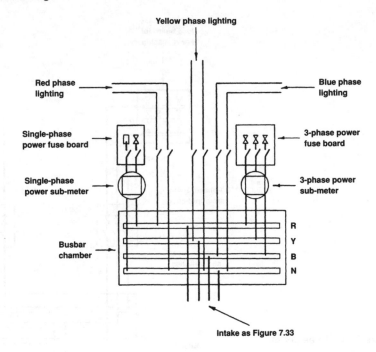

Figure 7.35 Factory supply incorporating power sub-meters. *Note:* Earth continuity through conduit and trunking

busbars in trunking will usually be favoured as shown in Figure 7.36, where single phases supply individual flats or business units in a high-rise block. Figure 7.37 emphasises the need for fire barriers as the trunking passes through compartment floors.

Extremely large and high-rise buildings

In these buildings, typical of city centre department stores and office blocks, the diagrammatic layout in Figure 7.38 indicates the combination of high current demand busbar distribution at low level, with various rising busbar chambers servicing each floor. Other three-phase fuse boards and panels are included for air-conditioning, boiler plant, lifts, etc. with automatic switch-over to emergency electricity generation.

Temporary supplies to building sites

An electricity supply to a building site is an essential service. It is required to operate plant ranging from low-voltage power tools to 400 V, three-phase hoists and cranes. A pre-contract meeting with the local electricity authority will establish the type of supply necessary, the total load required and a date for provision. The authority will provide an incoming site assembly containing their switchgear, fuses and meters, but thereafter the main contractor will

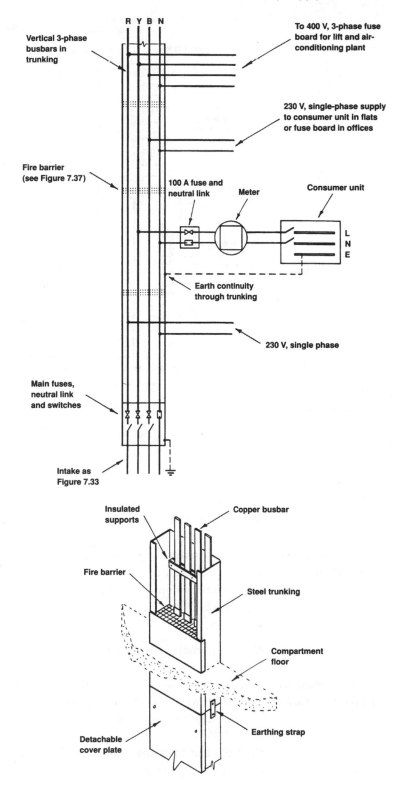

Figure 7.36 Distribution
for high-rise flats or offices

Figure 7.37 Vertical
busbar trunking

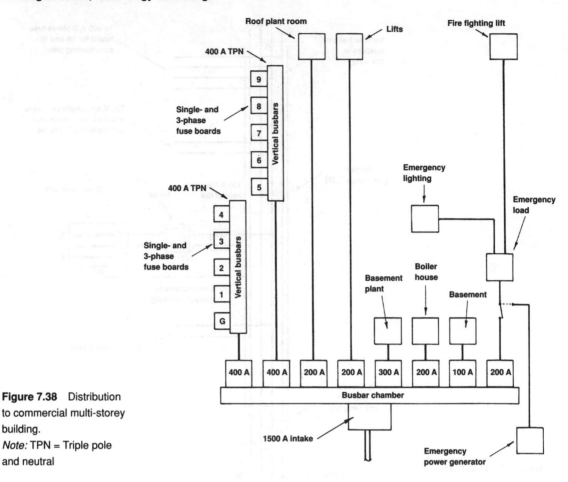

Figure 7.38 Distribution to commercial multi-storey building.
Note: TPN = Triple pole and neutral

have to subcontract the work to specialists. The equipment necessary follows the distribution shown in Figure 7.39 and includes the following:

Incoming site assembly (ISA)

- 400 V, three-phase and neutral cable
- fuses or circuit breakers
- current transformers, a single outgoing circuit of 300, 200 or 100 A
- switchgear
- meters.

Note: Main contractor to provide suitable damp-free housing.

Main distribution assembly (MDA)

This contains three-phase and single-phase distribution equipment and lockable circuit switchgear. Overload protection is by moulded case circuit breakers on each outgoing supply. MCCBs respond to greater prospective fault currents than MCBs.

Note: ISA and MDA are sometimes combined as one unit (ISDA) depending on the total requirements.

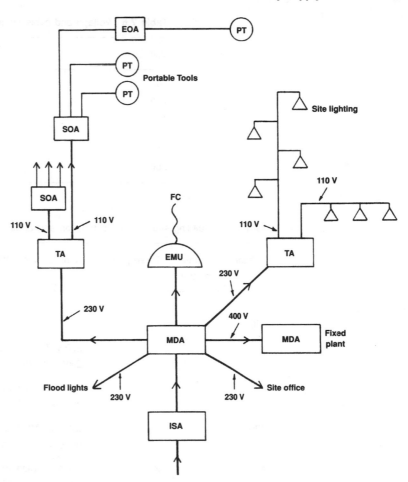

Figure 7.39 Temporary site distribution

| **Transformer assembly (TA)** | Several of these will be supplied from the MDA. Voltage transforms to 110 V and possibly 50 or 25 V in confined and damp situations (see Table 7.7). |

Transformer assembly (TA) Several of these will be supplied from the MDA. Voltage transforms to 110 V and possibly 50 or 25 V in confined and damp situations (see Table 7.7).

Earth monitor unit (EMU) This is used where flexible cables (FC) are needed to supply mains voltage to movable plant. In addition to the main earth conductor, a separate pilot conductor is incorporated to carry a very low-voltage current between plant and EMU and the earth conductor. If the current is interrupted, the monitoring unit detects it and trips the circuit.

Socket outlet assembly (SOA) Supplied at 110 V, 32 A and provided with switchgear and MCBs for connection of final subcircuits. Up to eight 16 A double pole socket outlets to portable tools.

Extension outlet assembly (EOA) Plugs into SOA and incorporates up to four 16 A double pole socket outlets. *Note:* Distribution units to satisfy the requirements of BS 4363 'Specification for distribution assemblies . . . on . . . sites'.

Table 7.6 Voltage and cable colour codes

Voltage	Colour
25	Violet
50	White
110	Yellow
230	Blue
400	Red
500–650	Black

Table 7.7 Building site voltage distribution and applications

Voltage	Single phase	Three phase	Application/situation
25	✓		} TA to restricted, damp and dangerous places
50	✓		
110	✓		230 V TA to hand and portable tools up to 2 kW and to site lighting
110		✓	400 V TA to movable site machinery, such as pumps, small concrete mixers and vibrators
230	✓		ISDA or MDA to site administration centre, general offices and floodlights
400		✓	ISDA or MDA to plant temporarily fixed and rated over 3.75 kW, e.g. cranes, hoists, large concrete mixers and compressors

Plugs, sockets and cables

In addition to variations in plug and socket profiles, defined in BS EN 60309–2 'Specification for industrial plugs, sockets . . .', colour identification is also recommended. Table 7.6 indicates the adopted coding for accessories and cables.

Central heating wiring diagrams

Manufacturers of boilers and central heating ancillaries produce design manuals to complement their products. Within these can be found several diagrams to suit the degree of automation required and the interrelationship of thermostats, time controls, motorised valves, etc. The possibilities are extensive, but for illustration purposes Figures 7.40 and 7.41 are shown to supplement

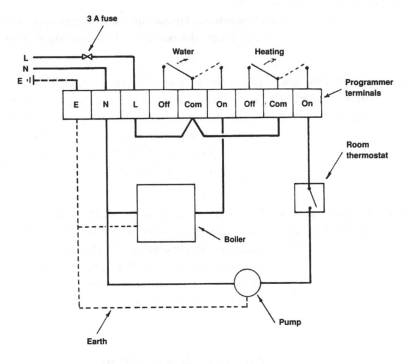

Figure 7.40 Wiring diagram for gravity primaries and pumped heating

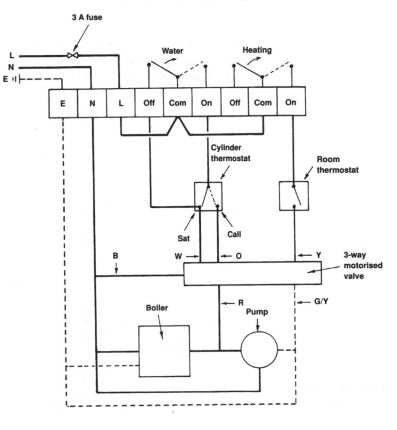

Figure 7.41 Fully pumped system.
Key: Com = common
Sat = hot water satisfied.
Sheathed cables:
W = white
O = orange
Y = yellow
B = blue
R = red
G/Y = green and yellow

the schematic layouts shown in Figure 3.24 (partial control of hot water) and Figure 3.26 (thermostatic control with three-port diverter) respectively.

Gravity primaries and pumped heating (Figures 3.24 and 7.40)

This is the simplest of systems, relying on gravity or convected circulation of hot water through the primary pipework to the calorifier and pumped circulation to radiators. Depending on the controls selected, many (including Figure 7.40) will only permit the central heating to function while the hot water is programmed, i.e. hot water functions without central heating, but selection of central heating also selects hot water.

Fully pumped system (Figures 3.26 and 7.41)

This system allows independent selection of hot water and central heating, promoting fuel efficiency and a rapid response to demand, as boiler and pump operate simultaneously. The mid-position or three-way motorised valve permits pumped water to be delivered to the calorifier, radiators or both. It can also be wired to provide priority for heating or hot water, i.e. one must be satisfied before the other is served.

Extra low-voltage lighting

This is a fairly recent development, gaining popularity for shop display lighting and office illumination. The system operates at only 12 V AC, via a 230 V single-phase transformer and fused splitter unit shown in Figure 7.42. The

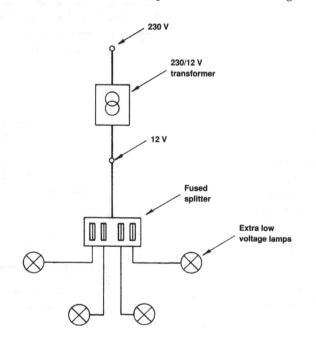

Figure 7.42 Extra low voltage lighting

combination of low voltage and high performance of the 50 W tungsten halogen dichroic lamps, offers several advantages over conventional lighting:

- low heat emission
- excellent colour rendering
- very low running costs.

Lamps are very sensitive to variations in voltage, so where several are required a splitter unit is used to avoid variable lead lengths running parallel from the transformer. Also, cable sizing is critical to maintain lamp efficiency; for example, a 50 W lamp at 12 V supply,

$$\text{Amps} = \frac{\text{Watts}}{\text{Volts}} = \frac{50}{12} = 4.17$$

Reference to Table 7.2 indicates that 1.00 mm^2 cable can convey up to 15.5 A, but the voltage drop of 44 mV is high. For a 4 m cable:

$$\frac{44 \times 4.17 \times 4}{1000} = 0.73 \text{ V} \quad \text{(6 per cent volt drop)}$$

Note: A drop of 0.7 V will reduce illumination by about 30 per cent. Overvoltage, too, must be avoided as this will shorten lamp life considerably.

Installation testing

Completed installations are subjected to an extensive programme of tests, defined in the IEE Regulations. Testing is a combination of visual inspection and procedures conducted with instrumentation (ohmmeter), and includes the following:

- polarity
- insulation
- continuity.

Polarity test   This is essential to ensure that switches, fuses and circuit breakers are connected in phase and not the neutral. Figure 7.43(a) shows the effect of malpractice and (b) the connection of an ohmmeter (bell circuit or test lamp also acceptable) across the live bar (disconnected!) and live terminals at switches, sockets and lamp holders. Operation of switches will respond on the ohmmeter.

Insulation test   This is to test the resistance between both or either poles of supply (phase and neutral) and earth. Ohmmeter test leads connect at the consumer unit as shown in Figure 7.44 and if resistance is seen to be low, current leakage will occur through a short circuit, causing energy wastage and a potential fire hazard.

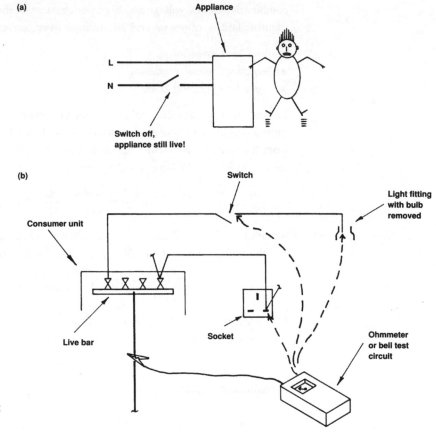

(a)

Appliance

L

N

Switch off,
appliance still live!

(b)

Switch

Light fitting
with bulb
removed

Consumer unit

Live bar

Socket

Ohmmeter
or bell test
circuit

Figure 7.43   Polarity test:
(a) malpractice; (b) test

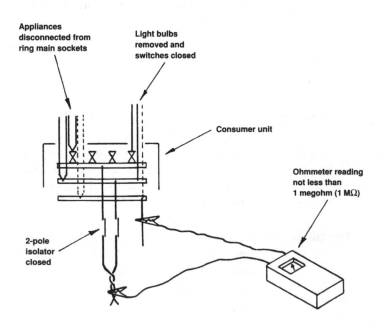

Appliances
disconnected from
ring main sockets

Light bulbs
removed and
switches closed

Consumer unit

Ohmmeter reading
not less than
1 megohm (1 MΩ)

2-pole
isolator
closed

Figure 7.44   Insulation
test

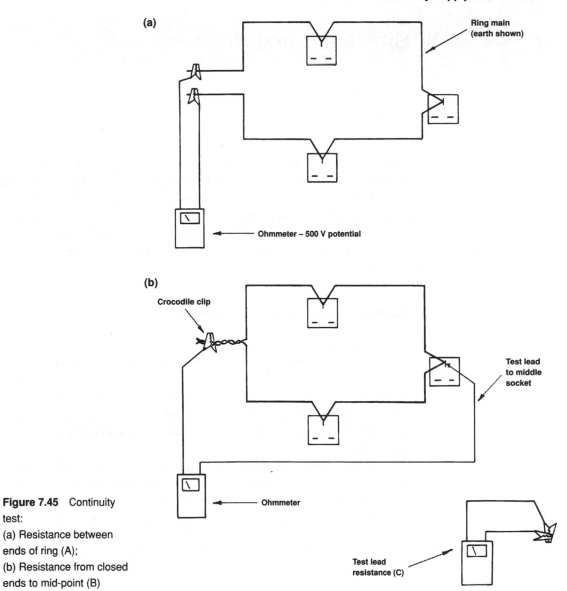

Figure 7.45 Continuity test:
(a) Resistance between ends of ring (A);
(b) Resistance from closed ends to mid-point (B)

Continuity test This, as the name suggests, is to ensure the integrity of the phase, neutral and earth conductors in a ring main. The procedure for each conductor is:

1. Record resistance between ends of the ring (A), see Figure 7.45(a).
2. Record resistance from closed ends to mid-point of ring (B), see Figure 7.45(b).
3. Record test lead resistance (C).
4. Check circuit integrity by formula,

$$\frac{A}{4} \text{ approx.} = B - C$$

8 Sanitation and drainage

Design and installation recommendations for sanitation and drainage are found in BS 5572 and enforced through the Building Regulations, Approved Document H. These set out acceptable standards, practice and design to minimise the risk to health and safety of building users. Drainage is a combination of above- and below-ground systems, known respectively as sanitation and foul drainage. Surface or rainwater collection and disposal can be considered separately.

Sanitation components and systems

All sanitary fitments discharging into a system of drainage, must be fitted with a water seal device or trap, to prevent the ingress of foul air from the drain and sewer. Figure 8.1 shows a conventional tubular 'S' or 'P' trap with the more compact bottle trap variation. For most applications, depth of water seal is 75 mm as indicated in Table 8.1, but where waste discharges are open, i.e. to a gully or hopper, trailing off will occur and 40 mm depth of seal traps may be used.

In addition to preventing foul air entering a building, a trap will also retain debris that could otherwise block the discharge pipe. For this purpose a cleaning eye or detachable access must be incorporated in the design. If the trap is integral, such as that on a WC pan, the appliance should be removable.

Loss of seal Under normal working conditions and test procedures, the depth of water seal in a trap must never be less than 25 mm. However, poor design, bad

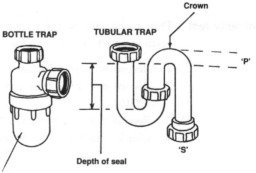

Figure 8.1 Traps

208

Table 8.1 Sanitation trap and waste data

Appliance	Trap/waste nominal bore (mm)	Trap seal (mm)
Basin	32	75
Bidet	32	75
Sink	40	75
Bath	40	75
Shower	40	75
Washing machine	40	75
Dishwasher	40	75
Urinal bowl	40	75
Food waste disposal unit	40	75
Sanitary towel macerator	40	75
Industrial food waste disposal unit	50	75
Urinal stall (1–6 person)	65	50
WC pan (siphonic)	75	50*
WC pan (washdown)	100	50*

* Trap integral.

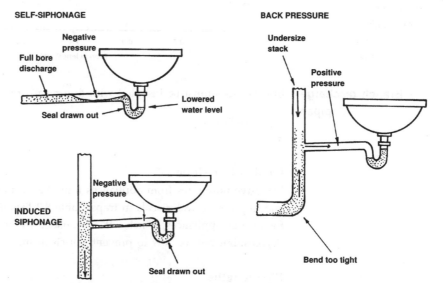

Figure 8.2 Seal loss in traps

workmanship or a defective fitting could contribute to seal loss. The most likely causes are self-siphonage or induced siphonage, or possibly air pressure fluctuations due to compression at the base of the stack. Siphonage will occur if the waste pipes are too long, too steep or too small in diameter. Also, incorrect disposition of appliances can lead to the waste from one inducing seal loss from the trap of another. Compression results from an undersized stack or a bottom bend too tight in radius, as illustrated in Figure 8.2.

Table 8.2 Sanitation design data

Appliance	Capacity (l)	Max. flow (l/s)	Branch slope (mm/m)
Basin	6	0.6	20–120 (Figure 8.3)
Urinal	4.5	0.15	
Sink	23	0.9	
Bath	80	1.1	18–90
Washing machine	180	0.7	
Shower	—	0.1	
WC	9	2.3	9 min.

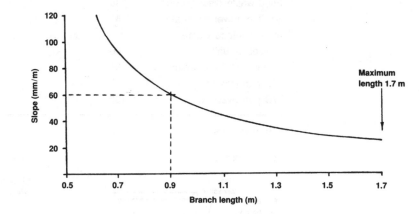

Figure 8.3 Basin waste pipe limitations. For example, a 0.9 m waste pipe has a corresponding slope of 60 mm per m length, i.e. 54 mm overall

Branch discharge pipes

Branch pipes must be installed with due regard to:

- gradient, and
- length.

Gradient or slope

Excessive flow rates from appliances must be avoided to prevent siphonage and to preserve water seals in traps. Table 8.2 lists the maximum acceptable for various appliances, which will be satisfied by limiting gradients to a few degrees, but not too low to prevent self-cleansing.

Pipe lengths

Basin and bidet branch pipes of 32 mm nominal bore, must not exceed 1.7 m as shown in Figures 8.3 and 8.4, with due regard for the gradient. Sinks, baths, etc. with 40 mm nominal bore waste pipes can extend up to 3 m as shown in Figure 8.4. WC pans are unlikely to siphon with normal usage and will operate successfully with branch pipes up to 6 m from the stack. In practical interests, it is usual to limit the branch length to as little as possible. Where grouping of appliances permits, the resulting single stack system shown in Figure 8.5 will satisfy most domestic situations.

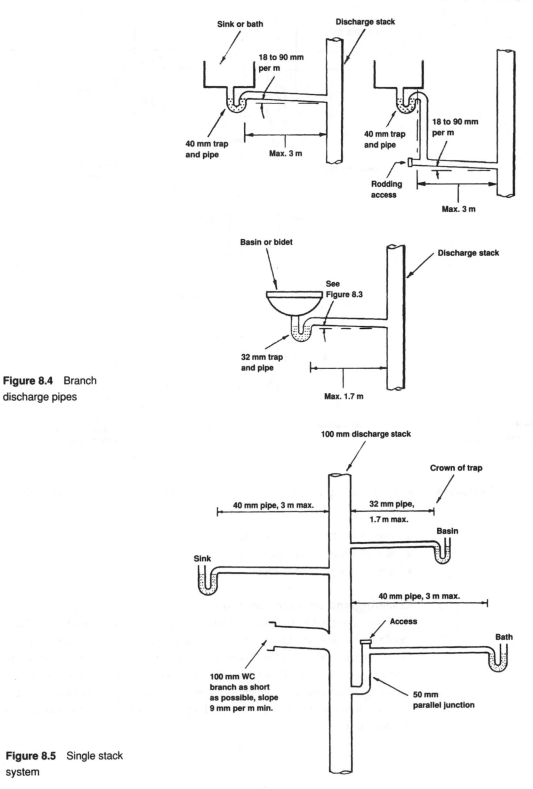

Figure 8.4 Branch discharge pipes

Figure 8.5 Single stack system

211

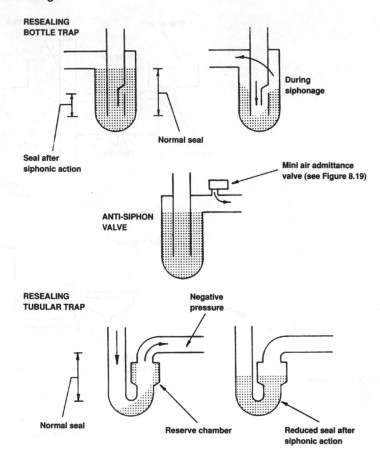

Figure 8.6 Anti-siphon and resealing traps

Exceptions If it proves impractical to satisfy the preceding requirements, a number of other options exist:

- use of resealing/anti-vacuum traps
- use of larger branch pipes
- ventilation of branch pipes.

Resealing and anti-vacuum traps

These are designed to maintain a water seal under extreme conditions. They should be considered a last resort, as they have the disadvantages of noise and the need for periodic attention, which is easily overlooked. If they are not maintained regularly, they will malfunction and could prove a health risk. Figure 8.6 shows some examples which work on the principle of a reserve chamber retaining water and the alternative anti-siphon attachment to the crown of a standard trap.

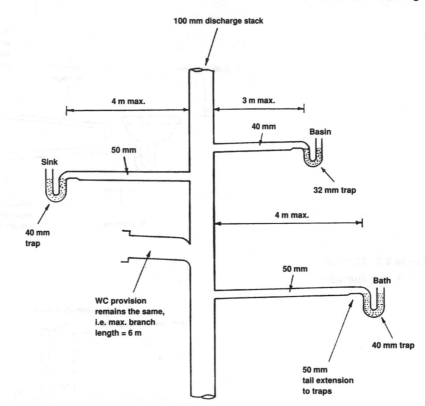

100 mm discharge stack

4 m max.

3 m max.

40 mm

Basin

50 mm

Sink

32 mm trap

40 mm
trap

4 m max.

50 mm

Bath

WC provision
remains the same,
i.e. max. branch
length = 6 m

40 mm trap

50 mm
tail extension
to traps

Figure 8.7 Extended branch discharge pipes. *Note:* For 40 and 50 mm waste pipes, the slope is between 18 and 90 mm per m

Larger branch discharge pipes

Enlarging the waste pipe diameter will provide increased flexibility for the design and installation of sanitary fitments. Figure 8.7 shows the potential of increasing basin and bidet wastes to 40 mm and sinks, showers, etc. to 50 mm. With all appliances, the trap diameter remains the same as original and the tail is provided with a 50 mm extension before the increase in diameter.

Ventilation branch pipes and stacks

If the occasional appliance waste pipe exceeds the design parameters, it may be ventilated to atmosphere by a vented branch as shown in Figure 8.8. Where it is impractical to provide short branches or several appliances share the same waste pipe, the whole system is vented as shown in Figure 8.9. The vent stack should be at least half the diameter of the wet stack, and the branch vents at least two-thirds of the diameter of the waste pipes served. Figure 8.10 shows a variation on the single stack domestic system which is appropriate for high-rise flats, where variable use of appliances could disturb water seals. The cross-connection between stacks will relieve negative pressure.

Cross-flow prevention Branch pipes must connect to a stack in such a way that the discharge from one appliance cannot back up the waste pipe of another. This necessitates a

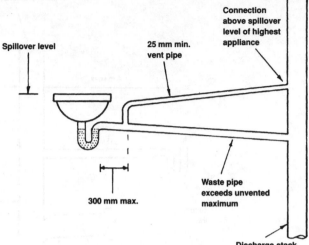

Figure 8.8 Modified branch discharge pipe

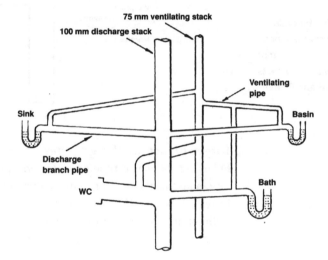

Figure 8.9 Fully ventilated system

stagger or offset of branch pipes sufficient to avoid interaction. The amount varies with pipe diameter and application; details are shown in Figure 8.11. Opposing bath and WC branches are the most affected and manufacturers have devised bosses and manifolds to overcome the problem. An example is shown in Figure 8.11.

Ground floor appliances

Ground floor waste pipes some distance from the stack can discharge into a gully or stub stack. The gully connection is through the surface grating or by purpose-made back or side inlet as shown in Figure 8.12. Ground floor WCs may only discharge directly into a drain, but this too can be via a stub stack, provided the drain is ventilated through another stack and limitations shown in Figure 8.13 are observed.

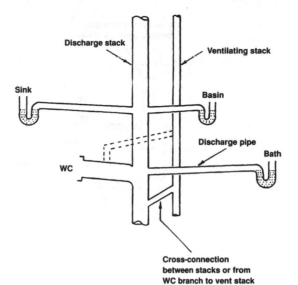

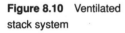

Figure 8.10 Ventilated stack system

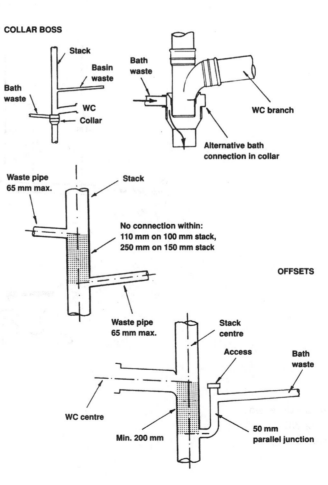

Figure 8.11 Prevention of cross-flow

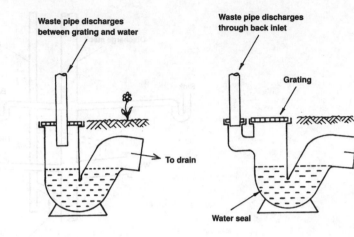

Waste pipe discharges
between grating and water

Waste pipe discharges
through back inlet

Grating

To drain

Water seal

Figure 8.12 Ground floor
waste pipe connection to
gullies

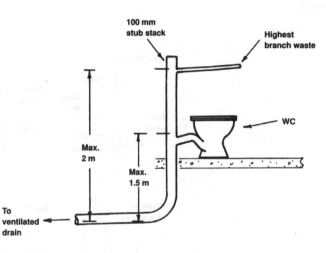

100 mm
stub stack

Highest
branch waste

WC

Max.
2 m

Max.
1.5 m

To
ventilated
drain

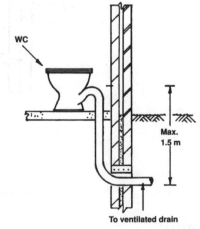

WC

Max.
1.5 m

To ventilated drain

Figure 8.13 Unvented
ground floor WC
connections to drain

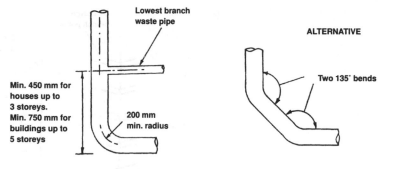

Figure 8.14 Stack base

Stack base connections

To avoid back pressure or compression at the bottom of the stack, a large radius bend must be provided. This should be at least 200 mm to the pipe centre or fabricated from two 135° bends shown in Figure 8.14. Also shown are the minimum distances of 450 and 750 mm from stack base to the lowest branch in dwellings up to three storeys and buildings up to five storeys respectively.

In high-rise buildings over five storeys, the ground floor appliances must not connect to the common stack, as pressure fluctuations at and near the stack base can affect trap seal retention. Appliances must discharge directly to a drain or gully or have their own stack as shown in Figure 8.15. Very tall buildings exceeding 20 storeys have both ground and first floor appliances discharging into their own stack.

Common or shared branch discharge pipes

Where several appliances share a branch pipe, such as a range of WCs or basins in commercial or public premises, venting as shown in Figure 8.16 will be necessary to prevent pressure fluctuations. However, if the following criteria can be achieved, the vent pipes are unnecessary:

- maximum of eight WCs connected to a 100 mm branch, not exceeding 15 m in length and graded between 9 and 90 mm per metre run
- maximum of five urinal bowls (six stalls) connected to a 50 mm (65 mm) diameter branch pipe as short as practical and graded between 18 and 90 mm per metre run
- maximum of four wash basins connected to a 50 mm diameter branch, not exceeding 4 m straight length, graded between 18 and 45 mm per metre length.

Offsets

Stacks should be vertical, but occasionally bends and offsets are necessary to avoid structural obstacles. If possible they should be avoided as they impede flow and may contribute to seal loss in nearby traps. If unavoidable, they must be of large radius and no appliance branch connection is permitted within 750 mm. If the wet stack has a complementary vent stack, it should be cross-vented as shown in Figure 8.17. Bends and offsets in the dry part of the stack, i.e. above the highest branch, are acceptable and frequently used where an

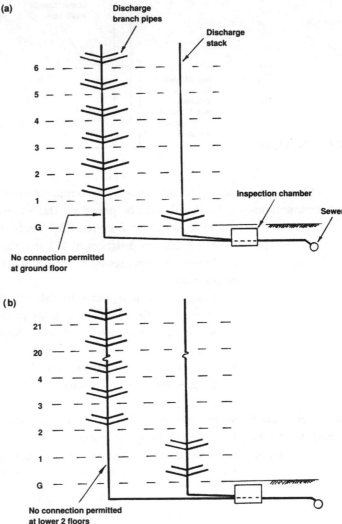

(a)

Discharge branch pipes

Discharge stack

No connection permitted at ground floor

Inspection chamber

Sewer

(b)

Figure 8.15 Discharge stacks in high-rise buildings:
(a) 5–20 storeys;
(b) over 20 storeys

No connection permitted at lower 2 floors

external stack detours around the eaves. External stacks are permitted for buildings up to three storeys, thereafter the stack must be within the building.

Ventilation   Stacks are conventionally vented with the top of the pipe projecting above the roof line and finished with a cage or perforated dome as shown in Figure 8.17. Penetration of the tiling is weathered with a lead slate or proprietary variation. In one- or two-storey houses, the dry part of a ventilation stack may reduce to 75 mm diameter.

Through ventilation can be achieved by either of the two methods shown in Figure 8.18. The traditional pre-1950s system uses an interceptor trap to separate the drain from the sewer. These are susceptible to blockage, but are occasionally still specified where it is necessary to prevent rodent penetration from the main sewer. Stacks contemporarily vented may be fitted with an

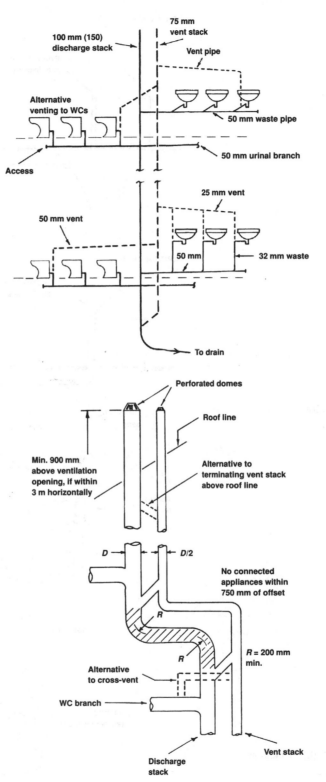

Figure 8.16 Stack
ventilation in large
buildings

Figure 8.17 Stack
offsets and termination

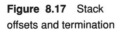

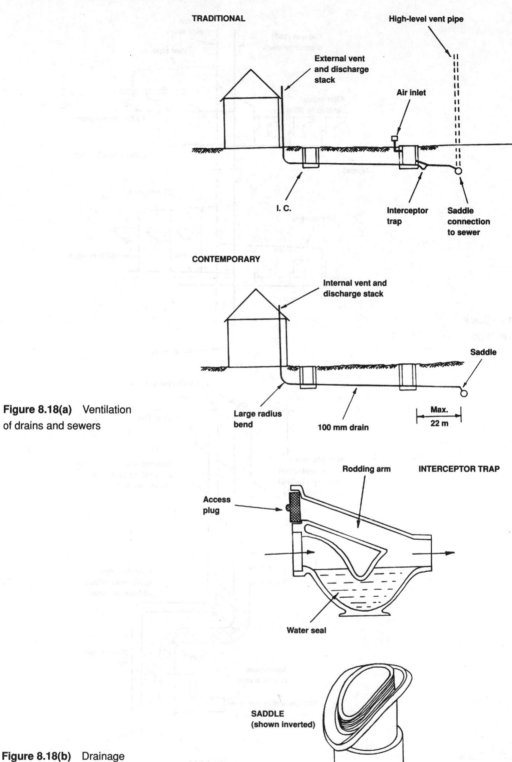

TRADITIONAL

High-level vent pipe

External vent
and discharge
stack

Air inlet

I. C.

Interceptor
trap

Saddle
connection
to sewer

CONTEMPORARY

Internal vent and
discharge stack

Saddle

Figure 8.18(a) Ventilation
of drains and sewers

Large radius
bend

100 mm drain

Max.
22 m

Rodding arm   **INTERCEPTOR TRAP**

Access
plug

Water seal

SADDLE
(shown inverted)

Figure 8.18(b) Drainage
accessories

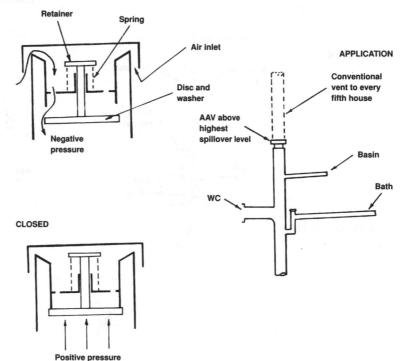

Figure 8.19 Air admittance valve (AAV)

air admittance valve, terminating inside the roof space and not less than the highest water level of an appliance connected to the stack. Figure 8.19 shows the application and operation of the device. It contains a diaphragm or disc which opens to allow fresh air into the stack when appliances discharge. A spring returns the diaphragm to seal and prevent foul air escaping while the stack is not functioning. Up to four consecutive dwellings of no more than three storeys can have an air admittance valve; the stack in the fifth dwelling must be conventionally vented to prevent a build-up of foul air in the drain and sewer.

Access   Accessibility to all traps, discharge branch pipes and stacks is necessary for system maintenance. Modern traps are easily detached to provide access into branch pipework, but if this is not convenient a rodding eye can be provided at the end of the pipe. Stacks can be rodded from the top and through access plates located midway between floors as shown in Figure 8.20. These should be no further than three storeys apart and where pipework is enclosed, doors or removable sections are required.

Airtightness   To satisfy the Building Regulations, an air or smoke test producing a pressure of at least 38 mm water gauge must be applied to completed sanitary pipework. Smoke testing is not recommended for use with uPVC pipes and is less easily

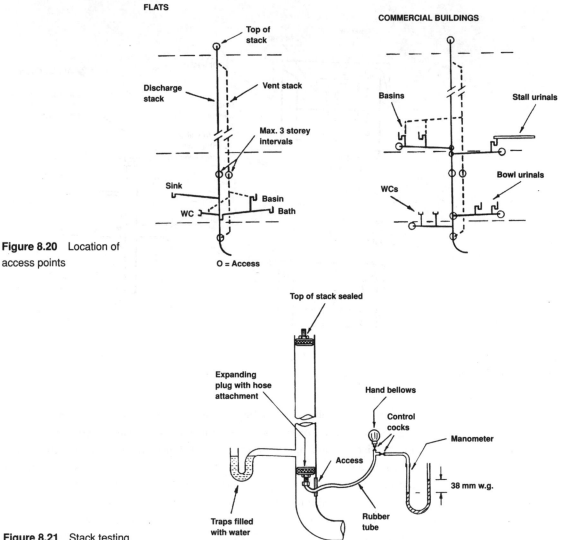

Figure 8.20 Location of access points

Figure 8.21 Stack testing

applied than the air test shown in Figure 8.21. All traps are charged with water, the stack top sealed and test plug applied at the lowest access plate. If this is inconvenient, access is possible at the highest inspection chamber or by plugging the inspection chamber and passing the test flexible tube through the water seal of a WC pan. The pressure must be retained for 3 minutes and during this time every trap must maintain at least 25 mm of water seal.

Sanitation system design

As a general guide, a 100 mm diameter stack will satisfy most requirements for buildings up to six storeys. Thereafter and for complex systems, formulae or unit allocation methods can be deployed to calculate appropriate stack size.

Formulae

Prior to using a stack design formula, it is necessary to assess the simultaneous demand factor. The following formula considers the number of appliances discharging simultaneously compared with the total number fitted:

$$m = np + 1.8(2np[1 - p])^{0.5}$$

where m = number of appliances discharging simultaneously
n = number of appliances installed
$p = t/T$
t = time in seconds for discharging
T = time between usage.

For example, a commercial building of 8 floors has 8 WCs, 10 basins, 4 sinks and 5 urinals at each level. Calculate:

- the simultaneous demand factor, and
- the volume flow rate in litres per second.

The total number of appliances is 27 per floor, multiplied by 8 floors = 216 fittings.

Using 10 s (average) for each appliance to discharge its contents and 500 s intervals between use:

$$p = \frac{t}{T} = \frac{10}{500} = 0.02$$

therefore,

$$m = (216 \times 0.02) + 1.8(2 \times 216 \times 0.02[1 - 0.02])^{0.5}$$

$$= 4.32 + 1.8(8.46)^{0.5}$$

$$= 4.32 + 5.24 = 9.56$$

The simultaneous demand factor is

$$\frac{m}{n} = \frac{9.56}{216} = 0.044 \text{ or } 4.4 \text{ per cent}$$

With reference to the flow rates in Table 8.2,

$$8 \text{ WCs} \times 2.3 \text{ l/s} = 18.40 \text{ l/s}$$

$$10 \text{ basins} \times 0.6 \text{ l/s} = 6.00 \text{ l/s}$$

$$4 \text{ sinks} \times 0.9 \text{ l/s} = 3.60 \text{ l/s}$$

$$5 \text{ urinals} \times 0.15 \text{ l/s} = \underline{0.75 \text{ l/s}}$$

$$28.75 \text{ l/s} \times 8 \text{ floors} = 230 \text{ l/s}$$

Allowing 4.4 per cent simultaneous demand,

$$\frac{230}{1} \times \frac{4.4}{100} = 10.12 \text{ l/s}$$

Stack diameter can be obtained from the following formula:

$$q = K\sqrt[3]{d^8}$$

where q = quantity discharge (l/s)

K = constant of 0.000 032

d = diameter of stack (mm).

Transposing,

$$d = \sqrt[8]{\left(\frac{q}{0.000\ 032}\right)^3}$$

$$= \sqrt[8]{\left(\frac{10.12}{0.000\ 032}\right)^3}$$

$$= 115$$

i.e. specify a 150 mm nominal bore stack.

Discharge units – BS 5572

An alternative approach to stack design is based on discharge units, which represent the load-producing properties of appliances. Table 8.3 lists the values for various applications and Table 8.4 provides the stack diameter for approximate totals of discharge units.

Table 8.3 Discharge unit values

Appliance	Application	Unit value
WC	Domestic	7
	Commercial	14
	Congested/public	28
Basin	Domestic	1
	Commercial	3
	Congested/public	6
Bath	Domestic	7
	Commercial	18
Sink	Domestic	6
	Commercial	14
	Congested/public	27
Shower	Domestic	1
	Commercial	2
Urinal	—	0.3
Washing machine	—	4
1 group of WC, bath and 1 or 2 basins	—	14

Using the preceding example, the total discharge unit (DU) value will be

$$8 \text{ WCs} \times 14 \text{ DUs} = 112 \text{ DUs}$$

$$10 \text{ basins} \times 3 \text{ DUs} = 30 \text{ DUs}$$

$$4 \text{ sinks} \times 14 \text{ DUs} = 56 \text{ DUs}$$

$$5 \text{ urinals} \times 0.3 \text{ DUs} = \underline{1.5 \text{ DUs}}$$

$$= 199.5 \text{ DUs}$$

Multiplied by 8 floors = 1596 DUs

From Table 8.4 it can be ascertained that a 150 mm nominal bore stack will be adequate.

Table 8.4 Discharge units and stack diameter

Nominal bore (mm)	Approximate no. of DUs
50*	10
65	60
75†	200
100‡	750
150	5500

* Min. diameter for stack serving urinals.
† Min. diameter for stack serving siphonic WCs.
‡ Min. diameter for stack serving washdown WCs.

If ventilating pipes are required, they cannot be sized by discharge units as they contain pressure variations, not hydraulic loadings. The following can be used as a general guide:

Branch or stack diameter	Ventilating pipe min. diameter
Up to 75 mm bore	$\frac{2}{3}$ bore (min. 25 mm)
Over 75 mm bore	$\frac{1}{2}$ bore

Underground foul and surface water drainage

Underground drainage pipe materials have varied, to include cast iron, asbestos, concrete, vitrified clay, pitch fibre and uPVC. Contemporary practice shows a preference for concrete in the larger sewers, i.e. pipes over 300 mm nominal bore and clayware or uPVC for the smaller tributary drains.

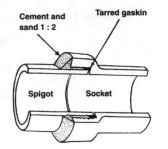

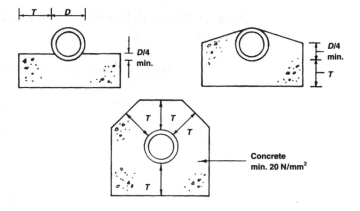

Figure 8.22 Traditional clayware spigot and socket joint

Figure 8.23 Traditional bedding of clay drains:
T = 150 mm min.;
D = diameter of pipe

Vitrified clay, BS EN 295

Clayware drain pipes are manufactured in 100, 150, 225 and 300 mm diameter nominal bore. Pipe lengths vary depending on size and manufacturer, but generally range from 1.3 to 1.6 m. The traditional spigot and socket, cement and sand jointed pipe shown in Figure 8.22 is now dated practice, but of course found in many existing drains. The difficulty of making a sound joint in wet trench conditions and the inability of this system to absorb movement, have rendered it obsolete. These drains are sometimes found with additional support from concrete bedding, haunching or surround as shown in Figure 8.23, which would be prohibitively expensive today.

Flexible jointing and bedding

The problem of pipe barrel failures and an awareness of ground movement potential led to the development of a modified spigot and socket joint. This is shown in Figure 8.24 containing polyester fairings to both spigot and socket and rubber sealing ring fitted to the recess in the spigot moulding. This permits sufficient angular movement and lineal draw without leakage, provided the pipes are bedded in a granular flexible medium. Some examples of this are shown in Figure 8.25. An alternative is plain-ended pipe joined with a polypropylene sleeve coupling, containing a sealing ring inside the rim at each end. This is also shown in Figure 8.24.

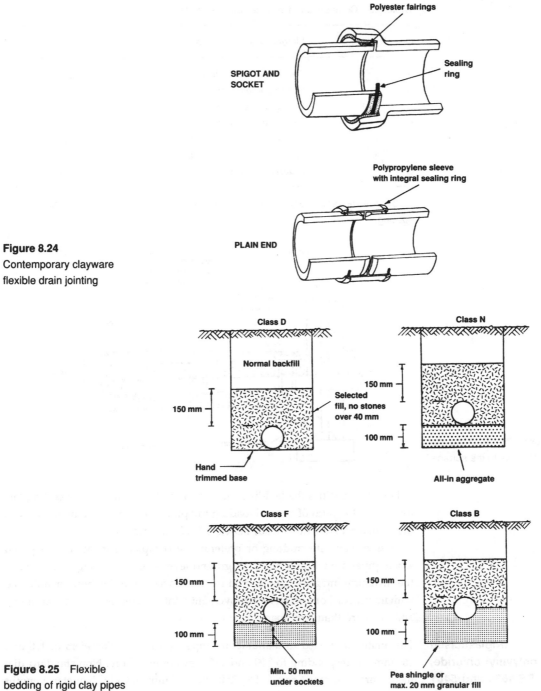

Figure 8.24
Contemporary clayware
flexible drain jointing

Figure 8.25 Flexible
bedding of rigid clay pipes

Table 8.5 Comparison of pipe bedding techniques

Class	Material and technique	Bedding factor
A	Reinforced concrete cradle*	3.4
A	Plain concrete cradle*	2.6
S	360˚ granular surround	2.2
B	180˚ granular support	1.9
F	Flat layer, single size granules	1.5
N	Flat layer, all in aggregate	1.1
D	Natural trench bottom	1.1

* 13 mm wide vertical gaps formed in concrete with fibre-board, at 5 m max. spacing.

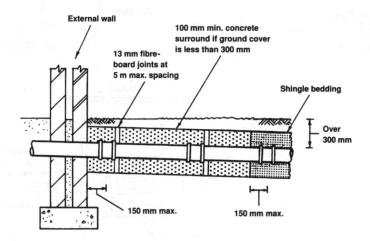

Figure 8.26 Bedding clay drains close to the surface

Table 8.5 specifies the bedding material and effectiveness as a bedding factor. This is the ratio of vertical loading the pipe will carry in comparison with the British Standard test load, i.e. class S is 2.2 times better.

Class A, concrete cradling or preferably a complete surround, is required where pipes are within 300 mm of ground level as shown in Figure 8.26. Here the structural integrity is preserved by the concrete and flexibility maintained with fibre-board or polystyrene movement joints, preferably at every coupling and no more than 5 m apart.

Unplasticised polyvinyl chloride, BS 4660 and 5481

This material is used to manufacture pipes in outside diameters of 110 and 160 mm, corresponding to 100 and 150 mm nominal bore. Non-standard diameters are also made to 82, 200, 250 and 315 mm outside diameter. Lengths are 3 and 6 m, easily cut with a hack-saw.

Jointing

Jointing of uPVC pipes can be by solvent welded (glued) spigot and socket, but this is usually restricted to sanitation pipework. Push-fit couplings with

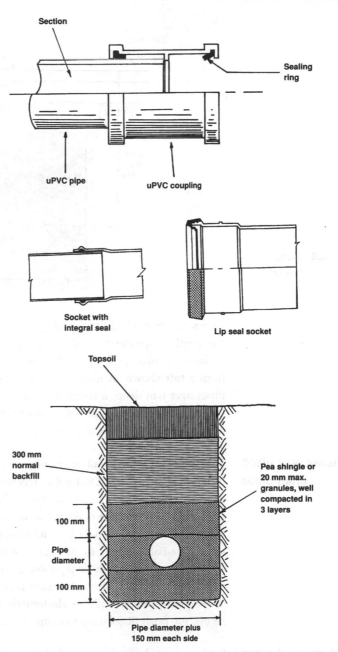

Figure 8.27 Jointing uPVC drain pipes

Figure 8.28 Bedding and backfilling plastic drainpipes

sealing rings are more appropriate for underground drainage as these will respond to movement without leaking. Both spigot and socket, and double socket or sleeve couplings are shown in Figure 8.27.

Bedding

uPVC pipes are sufficiently flexible to absorb modest ground movement, therefore bedding technique is not as critical as for rigid clayware. Figure 8.28

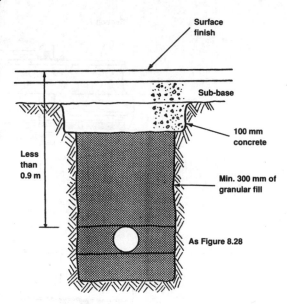

Figure 8.29 Bedding plastic drainpipes under roads

shows acceptable bedding which provides protection from deformity for 0.6–6 m depths. Pipe depths within 0.6 m of the pipe crown are acceptable if the surface is garden or footpath, but under roads or drives concrete protection from a raft shown in Figure 8.29 must be provided. Alternatively, and for pipes over 6 m deep, a complete surround of concrete is acceptable. During concreting the pipe is filled with water to prevent it floating and joints wrapped in polythene to prevent ingress of cement.

Structured wall uPVC drain pipes

Structured or profiled wall plastic pipes are a relatively new development, produced in standard sizes, but yet to receive BSI approval. As an interim measure, manufacturers are obtaining British Board of Agrément certification. The advantage to producers is economic use of material. In use the smooth bore is the same as conventional pipe with the external ribbing providing axial rigidity and radial strength to withstand superimposed loading. Not all water authorities will accept these pipes for use in sewers, as they have reservations about the material's strength under high-pressure jet cleaning. Figure 8.30 shows two profiles, with an elastomeric sealing ring located in the surface recess, for push-fit jointing to couplers and sockets.

Settlement of pipes

Settlement of drains is particularly important around and under a building. Flexibility and independence are important design considerations, as movement of the drain could impair the stability of adjacent buildings and conversely, settlement of a building could affect the efficiency and function of a drain.

Drains under a building

This should be avoided, but if there is no alternative, flexibly jointed pipes with at least 100 mm granular surround as shown in Figure 8.31 is acceptable.

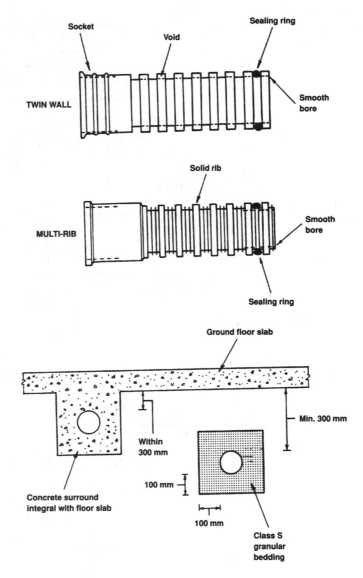

Figure 8.30 Ribbed uPVC pipes

Figure 8.31 Drains under a building

If the pipe crown is within 300 mm of the slab soffit, concrete encasement with the slab is required.

Drains through a wall or foundation

This is unavoidable as the drain needs access to the sanitation system, unless the stack is secured to an external wall. To protect the pipe against building settlement, provision of a void and/or rocker pipes as shown in Figure 8.32 will be necessary.

Drains below foundation level

Deep drains close to a building require protection from the superimposed ground loading. Figure 8.32 shows the necessary concrete surround and Figure 8.33 indicates the application for:

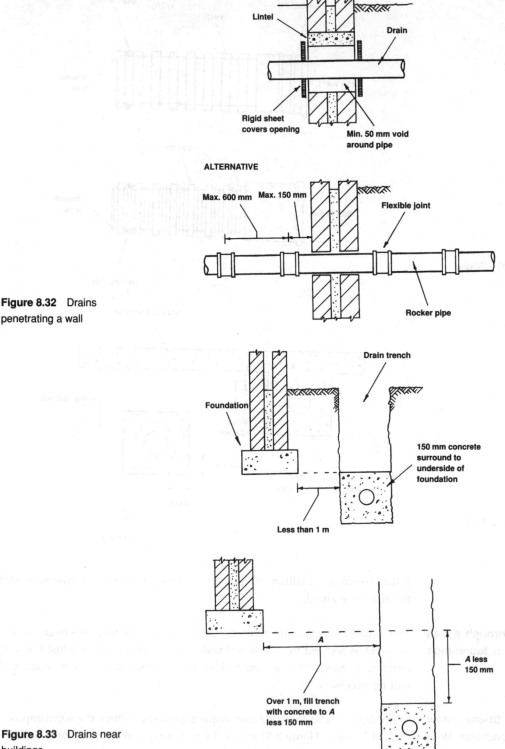

Figure 8.32 Drains penetrating a wall

Figure 8.33 Drains near buildings

- pipes under 1 m measured horizontally from the foundation, and
- pipes over 1 m.

Surcharging

If a drain is liable to surcharges from tides or backing up from the sewer, an anti-flooding interceptor trap should be fitted. This is a simple non-return mechanism where a hollow copper or plastic ball seals the drain from a flooding sewer.

Rodent control

Rats are well established in some sewer systems and remain elusive to controls. Where they are prevalent a conventional interceptor trap shown in Figure 8.18 will normally suffice. In the severest of circumstances, sealed drainage using access plates within inspection chambers may be necessary.

Drainage systems and components

Underground drainage can be divided into three categories:

1. **Drain** – a drain is the part of the system which serves one building only and is within the curtilage of that building. It includes all accessories such as inspection chambers, gullies, etc.
2. **Private sewer** – a private sewer is generally accepted as a drain which serves more than one building. It is within the curtilage of those buildings and is not usually adopted by the local authority.
3. **Public sewer** – a public sewer is a sewer or drain that has been vested in the local authority.

Exact parameters are impossible to define. Property deeds and local water and council authority sewer maps should be consulted for detailed clarification on individual responsibilities for maintenance and repair. Figures 8.34 and 8.35 show some differentiation with typical modern drainage layouts.

Combined system (Figure 8.34)

This comprises one pipe which receives both foul and surface water from the building. As the drain and sewer are shared, the installation costs are comparatively low, but sewage processing costs high, due to the extra volume of surface water. These systems are common in older development areas, but the water authorities rarely accept them now, due to the need for special overload facilities at sewage treatment works in event of storm-water surcharge.

Separate system (Figure 8.35)

This comprises two drains and two sewers, one to receive foul water and the other surface water. Site installation costs are higher than for a combined system, but sewage processing is much cheaper and controllable. Surface water can discharge into a stream, river or other convenient watercourse. In granular subsoils, a soakaway can be used.

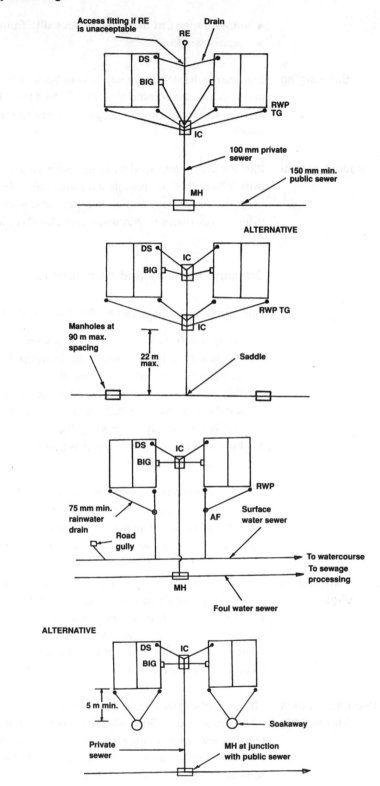

Figure 8.34 Combined systems of drainage.
Key:
DS = discharge stack;
BIG = back inlet gully;
RE = rodding eye;
IC = inspection chamber;
RWP = rainwater pipe;
MH = manhole;
TG = trapped gully;
AF = access fitting

Figure 8.35 Separate systems of drainage

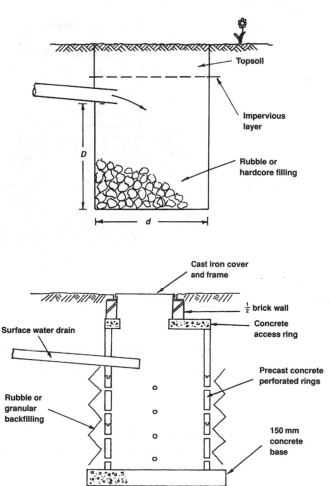

Figure 8.36 Simple filled soakaway.
D, the depth below pipe invert is approximately the same as *d*, the diameter

Figure 8.37 Hollow soakaway

Partially separate system

This is predominantly a separate system, but with local authority approval, the odd surface water downpipe from a garage roof for example, can discharge into the foul water drain if it is unreasonably difficult to connect this to the surface water drain.

Soakaways

A soakaway is shown in its simplest form in Figure 8.36. It is a pit filled with coarse rubble for collection and storage of storm-water, for subsequent dispersal into the subsoil. They are only acceptable in granular, free-draining subsoils and must be on land lower than the building to be drained and not closer than 5 m to the building. This acknowledges that water concentrations could undermine foundations. Figures 8.37 and 8.38 show acceptable soakaways for large buildings and estate drainage from several dwellings. Capacities and volumes for excavation can be determined from soil percolation tests and graphical

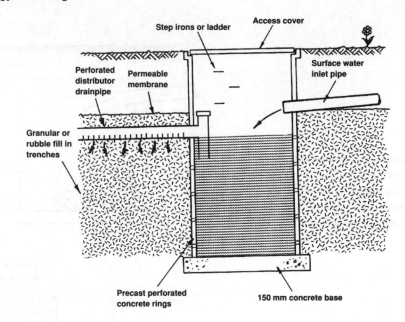

Step irons or ladder

Access cover

Surface water inlet pipe

Perforated distributor drainpipe

Permeable membrane

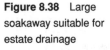

Granular or rubble fill in trenches

Precast perforated concrete rings

150 mm concrete base

Figure 8.38 Large soakaway suitable for estate drainage

analysis (see BRE *Digest* 365), but the following empirical formula provides sufficient guidance for soakaways serving small buildings:

$$C = \frac{AR}{3}$$

where C = capacity (m³)
A = area on plan to be drained (m²)
R = rainfall (m/h).

For example, a roof plan area of 80 m² and rainfall intensity of 75 mm/h:

$$C = \frac{80 \times 0.075}{3} = 2 \text{ m}^3$$

Layout of drains   For efficient discharge and minimal maintenance, the drainage layouts should incorporate the following features:

1. Simplicity
2. Straightness between access points with easy curves in or near chambers
3. Access points provided only where needed, i.e.
   (a) significant changes in gradient
   (b) significant changes in direction
   (c) change in drain diameter
   (d) junctions
   (e) at or near the head of each drain run
4. Adequate ventilation at the head of each drain, but not required if a branch drain is less than 6 m long (see also previous section on automatic air vents).

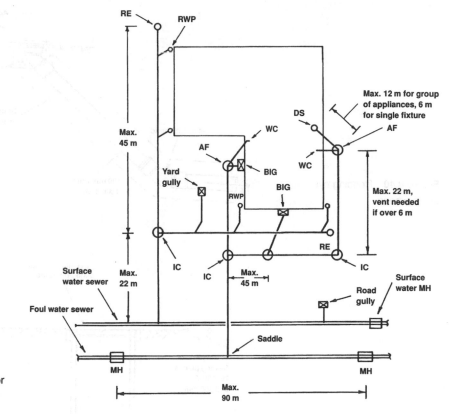

Figure 8.39 Access provision to domestic drains.
Note: See Figure 8.34 for key to abbreviations

In addition to Figures 8.34 and 8.35, Figure 8.39 shows provision for drain access to a typical detached estate house. It also incorporates Building Regulation dimensional requirements for location and spacing of means of access.

Means of access   Access to drains for inspection, testing and clearance of blockage can be achieved with:

- rodding eye (Figure 8.40)
- access fitting (Figure 8.41)
- inspection chamber (Figure 8.42), or
- manhole (Figures 8.43–8.45).

Rodding eye

These are a sealed surface extension to a drain, frequently located at the head of the drain as an alternative to an inspection chamber. They can also be deployed elsewhere if the drain run remains close to the surface, but have the disadvantage of permitting rodding in the direction of flow only.

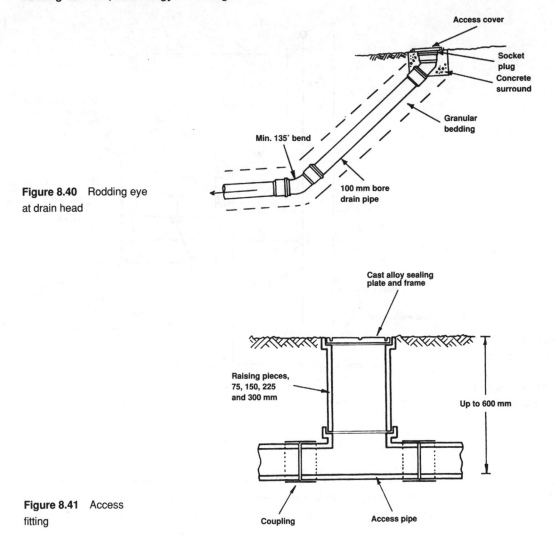

Figure 8.40 Rodding eye at drain head

Figure 8.41 Access fitting

Access fitting

These fittings are integral with the drain and provide similar access to a rodding eye, but with vertical access the drain can be rodded in both directions. They are suited for shallow depths up to 600 mm to invert level and are appropriate at or near the head of a drain.

Inspection chamber

Inspection chambers are an enlarged version of an access fitting with provision for a few junctions, sufficient for normal domestic use. Depth to invert is no more than 1 m, with surface access only. Construction may be from preformed plastic, precast concrete or built *in situ* masonry as shown in Figure 8.42.

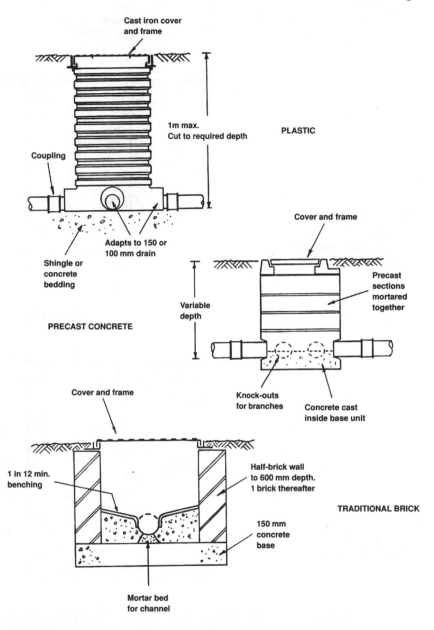

Figure 8.42 Inspection chambers

Manholes

These are masonry or precast concrete access chambers, with sufficient working space at drain level. Where construction and depths exceed 1 m to invert, step irons or a fixed ladder should be provided. Figures 8.43–8.45 show requirements for traditional masonry chambers, including shafts at depths exceeding 2.7 m.

A backdrop manhole is a variation which connects drains at significantly different levels. Figure 8.46 shows typical construction with vertical drop saving considerably on excavation costs.

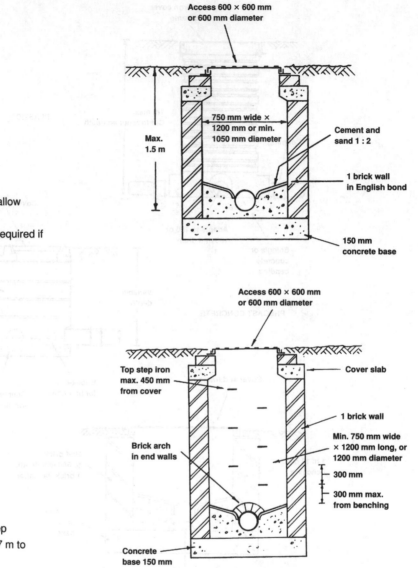

Figure 8.43 Shallow manhole.
Note: Step irons required if over 1 m deep

Access 600 × 600 mm or 600 mm diameter

750 mm wide × 1200 mm or min. 1050 mm diameter

Max. 1.5 m

Cement and sand 1 : 2

1 brick wall in English bond

150 mm concrete base

Figure 8.44 Deep manhole, up to 2.7 m to invert

Access 600 × 600 mm or 600 mm diameter

Top step iron max. 450 mm from cover

Cover slab

1 brick wall

Min. 750 mm wide × 1200 mm long, or 1200 mm diameter

300 mm

300 mm max. from benching

Brick arch in end walls

Concrete base 150 mm

Testing for watertightness

The two most convenient methods for testing drains are an air test and a water test. Either could be applied after drain laying and before backfilling, as discovery of leakage on completion of the building would be expensive to rectify.

Air test

The equipment consists of two drain stoppers, one with a rubber tube attachment and a manometer complete with another tube and hand bellows or a bicycle pump. The blank stopper is applied to the uppermost end of the drain run and the remaining equipment to the lower end as shown in Figure 8.47.

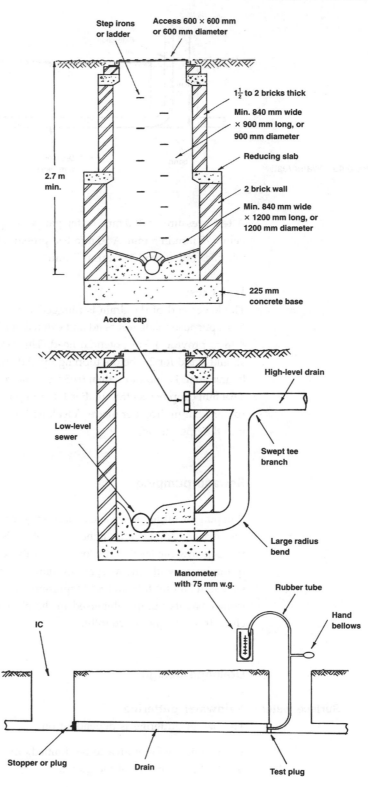

Step irons
or ladder

Access 600 × 600 mm
or 600 mm diameter

$1\frac{1}{2}$ to 2 bricks thick

Min. 840 mm wide
× 900 mm long, or
900 mm diameter

Reducing slab

2.7 m
min.

2 brick wall

Min. 840 mm wide
× 1200 mm long, or
1200 mm diameter

225 mm
concrete base

Figure 8.45 Deep
manhole with access shaft

Access cap

High-level drain

Low-level
sewer

Swept tee
branch

Large radius
bend

Figure 8.46 External
back drop manhole

Manometer
with 75 mm w.g.

Rubber tube

Hand
bellows

IC

Stopper or plug

Drain

Test plug

Figure 8.47 Air test

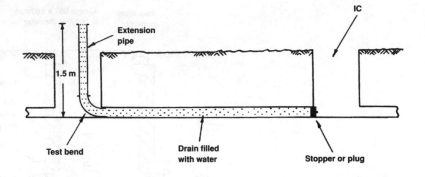

Figure 8.48 Water test

A test pressure of 100 mm water gauge is applied and this should not fall below 75 mm in 5 min. A 50 mm test pressure is also acceptable, provided the fall does not exceed 12 mm in 5 min.

Water test

The lower end of the drain is plugged with a blank stopper. The upper end has a purpose-made test bend and extension pipe attached as shown in Figure 8.48 to provide 1.5 m potential head. The extension is filled with water and should stand for 2 h before topping up. After a further 30 min leakage should be minimal, i.e. no more than 0.05 l per metre run of 100 mm drain equalling a 6.4 mm drop per metre run. For 150 mm pipes, no more than 0.08 l per metre run or 4.5 mm drop per metre. A leak will be apparent from observed seapage into the open trench.

Sewage pumping

Pumping or lifting of sewage is necessary from basement conveniences when the adjacent drain is at a higher level. It is also needed when a drainage system is below the level of the local authority sewer. Pumps should be installed in duplicate, with one as spare or stand-by if the duty pump fails. Systems shown in Figures 8.49 and 8.50 represent wet and dry sumps respectively. The former has its pump submersed in the effluent, but accessed by suspension chain from its gravity coupling.

Drainage design

Surface water   **Rainwater guttering**

The size of guttering is dependent on:

- the effective roof area to be drained, and
- the flow capacity of the gutter.

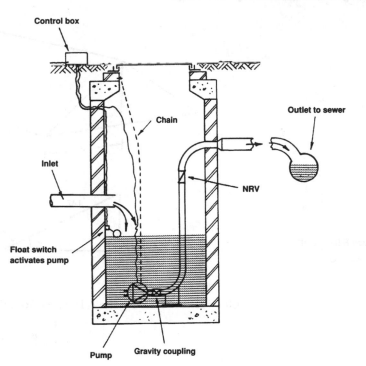

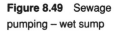

Figure 8.49 Sewage
pumping – wet sump

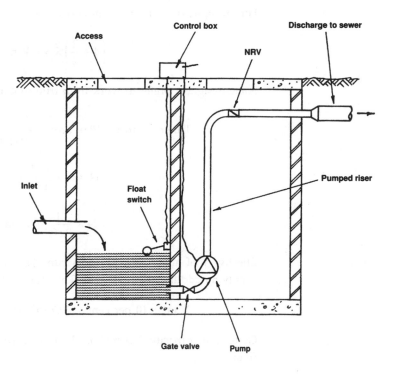

Figure 8.50 Sewage
pumping – dry sump

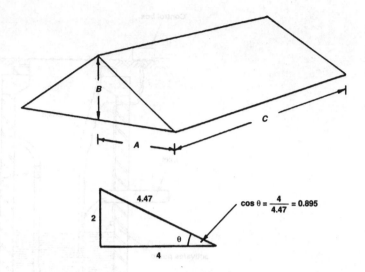

Figure 8.51 Gutter calculations

Effective roof area may be calculated from the following formula:

$$\left(A + \frac{B}{2}\right) \times C$$

For example, $A = 4$ m, $B = 2$ m and $C = 6$ m. So

$$\left(4 + \frac{2}{2}\right) \times 6 = 30 \text{ m}^2$$

Some designers prefer to work from true roof area:

$$\text{True roof area} = \text{Plan area} \times \frac{1}{\cos \text{pitch (Figure 8.51)}}$$

$$= (4 \times 6) \times \frac{1}{4/4.47}$$

$$= 24 \times \frac{1}{0.895} = 27 \text{ m}^2$$

Effective roof area, allowing for accumulation of water along the slope, is simply an additional 10 per cent. Therefore in this example

$$27 \text{ m}^2 + 10 \text{ per cent} = 29.7 \text{ m}^2 \quad (\text{round up to } 30 \text{ m}^2)$$

Gutter flow capacity depends on the rainfall runoff, which can be calculated from

$$Q = \frac{APR}{3600}$$

Table 8.6 Surface permeability factors

Surface type	Factor
Roofs (except thatch)	0.95
Asphalt	0.85–0.95
Concrete	0.85–0.95
Paving – mortared joints	0.75–0.85
– open joints	0.50–0.70
Blocks – open joints	0.40–0.50
Tarmacadam roads	0.25–0.60
Gravel drives	0.15–0.30
Grass (depends on soil)	0.05–0.25
Overgrown areas	0.01–0.20

Table 8.7 Gutter flow capacity

Gutter profile (mm)	End outlet flow capacity (l/s)		Centre outlet flow capacity (l/s)		Outlet (mm)
	Level	1 : 600	Level	1 : 600	
75 half-round	0.27	0.38	0.54	0.75	50
100 half-round	0.73	1.00	1.46	2.00	63
112 half-round	0.83	1.17	1.67	2.34	63
112 square	1.08	1.52	2.17	3.04	63
115 deep flow	2.05	2.87	3.97	5.56	63 or 75
130 ogee	1.80	2.56	3.60	5.13	75 or 89
150 half-round	2.30	3.23	4.60	6.46	75 or 89

where Q = quantity of water (l/s)

A = effective area to be drained (m²)

P = surface permeability (see Table 8.6)

R = rainfall (mm/h).

From the preceding example of 30 m² effective roof area and assumed rainfall of 75 mm/h,

$$Q = \frac{30 \times 0.95 \times 75}{3600}$$

$$= 0.6 \, l/s$$

From Table 8.7 it is possible to select gutter size, profile, outlet location and whether to fix the gutter level or to a slight fall. For this example, a 100 mm half-round gutter will suit all applications.

Table 8.8 Fluid flow in drains

Proportional depth	Velocity of flow (m/s)	Proportional depth	Velocity of flow (m/s)
0.125	0.770	0.600	1.460
0.200	0.970	0.666	1.500
0.250	1.060	0.700	1.510
0.300	1.140	0.750	1.526
0.333	1.210	0.800	1.535
0.400	1.290	0.900	1.520
0.500	1.390	Full	1.390

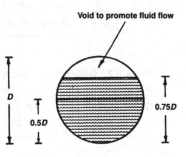

Figure 8.52 Proportional depth

Drainage of car parks, roads and large made-up areas

These situations use the same rainfall runoff formula as for roofing, but a rainfall intensity of 50 mm/h can be assumed sufficient, i.e.

$$Q = \frac{APR}{3600} = \text{litres per second}$$

For example, a car park 100 m × 100 m has a permeability factor of 0.8. Calculate the rainfall runoff and the main drain diameter.

Rainfall runoff:

$$Q = \frac{(100 \times 100) \times 0.8 \times 50}{3600}$$

$$= 111 \text{ l/s or } 0.111 \text{ m}^3/\text{s}$$

Pipe sizing formula:

$$Q = V \times A$$

where Q = quantity of water (m^3/s)

V = velocity of water flowing in drain (min. 0.75 m/s for self-cleansing)

A = area of water in the pipe (m^2).

The relationship between potential water velocity and depth of flow is listed in Table 8.8. Proportional depths are illustrated in Figure 8.52. Allowing 0.5 proportional depth or 50 per cent of the full bore to permit a reserve in event of extreme storms and a void to encourage flow in normal conditions, the velocity of 1.39 m/s will correspond.

Thus,

$$Q = V \times A$$

$$= 0.111 \text{ m}^3/\text{s}$$

$$V = 1.39 \text{ m/s}$$

$$A = \frac{\pi r^2}{2} \quad \text{i.e. half-bore}$$

Transposing,

$$A = \frac{Q}{V} = \frac{0.111}{1.390} = 0.08 \text{ m}^2$$

Therefore total area of pipe = $2 \times 0.08 \text{ m}^2 = 0.16 \text{ m}^2$.

$$A = \pi r^2$$

or

$$r = \sqrt{\frac{A}{\pi}}$$

$$= \sqrt{\frac{0.16}{\pi}} = 0.225 \text{ m}$$

Therefore diameter = $2 \times 0.225 = 0.450$ m or 450 mm.

Note: This will be supplied from several tributary drains, the diameter of which can be calculated in the same way from the reduced areas of car park served.

Foul water The size of foul water drains and sewers acceptable for up to 20 dwellings is 100 mm nominal bore, laid to a minimum gradient of 1 in 80. Above 20 and up to 150 dwellings, a 150 mm nominal bore pipe is sufficient provided the gradient is no less than 1 in 150. Situations outside these guidelines are sized similarly to surface water, except that runoff is established by summing discharge units (see Table 8.3) and converting these to a flow rate from Table 8.9.

Volume flow in foul drains can be designed up to 0.75 proportional depth, but where additional development is anticipated, it may be more prudent to reduce this to 0.5, i.e. half-bore. The following formulae provide varying values for Q, the flow rate (m³/s):

0.75 proportional depth	$Q = V \times (\pi r^2 - 0.617r^2)$
0.66 proportional depth	$Q = V \times (\pi r^2 - 0.839r^2)$
0.50 proportional depth	$Q = V \times 0.5 \, \pi r^2$
0.33 proportional depth	$Q = V \times 0.839r^2$
0.25 proportional depth	$Q = V \times 0.617r^2$

Table 8.9 Discharge units to flow rate

Discharge units	Approx. flow rate (l/s)	Discharge units	Approx. flow rate (l/s)
200	2.1	6 000	22.0
300	2.8	8 000	26.5
400	3.4	10 000	30.0
500	4.0	15 000	40.0
600	4.6	20 000	49.0
700	5.1	30 000	64.0
800	5.6	40 000	78.0
1 000	6.5	50 000	90.0
1 500	8.5	60 000	100.0
2 000	10.5	80 000	120.0
4 000	17.0	100 000	140.0

For example, calculate the pipe diameter for a sewer flowing 0.75 proportional depth, serving an estate producing 50 000 discharge units.

From Table 8.9, 50 000 DUs are approximately 90 l/s or 0.09 m³/s. From Table 8.8, 0.75 proportional depth generates a velocity of 1.526 m/s. Using the formula for 0.75 proportional depth:

$$Q = V \times (\pi r^2 - 0.617r^2)$$

$$0.09 = 1.526 \times (2.525r^2)$$

$$\frac{0.09}{1.526} = 2.525r^2$$

$$r^2 = \frac{0.059}{2.525}$$

$$r = 0.153$$

Therefore pipe diameter = 0.306 m, i.e. a 300 mm nominal bore pipe will just be sufficient.

Gradient or fall in drains and sewers

For simple situations such as branch drains to dwellings, the tables provided in Approved Document H to the Building Regulations are sufficient. Maguire's rule in imperial measures also has credibility in simple applications, i.e.

4 inch (100 mm) pipe inclined at 1 in 40,

6 inch (150 mm) pipe inclined at 1 in 60,

9 inch (225 mm) pipe inclined at 1 in 90, etc.

However, mathematical and scientific research proves that much lower gradients can successfully self-cleanse. Studies dating back to the eighteenth century

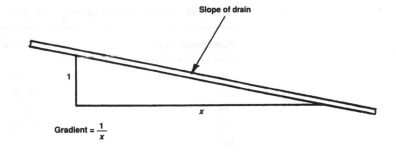

Figure 8.53 Gradient or slope expressed as a fraction

Gradient = $\dfrac{1}{x}$

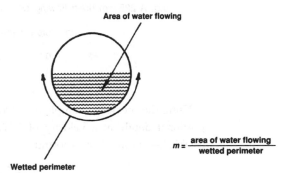

Figure 8.54 Hydraulic mean depth (m). *Note:* For full and half-full bore situations, m = pipe diameter/4. For other depth proportions, see Table 8.9

$m = \dfrac{\text{area of water flowing}}{\text{wetted perimeter}}$

reveal the Frenchman, Antoine Chezy's acknowledged formula for flow in sloping channels and pipes. Chezy's formula:

$$V = C\sqrt{m \times s}$$

where V = velocity of flow (m/s)
$\quad C$ = Chezy coefficient
$\quad m$ = hydraulic mean depth
$\quad s$ = slope as a fraction, $1/x$ (see Figure 8.53).

The expression $\sqrt{m \times s}$ is effectively acceleration and will vary with depth of water flowing and the gradient. While the gradient may be easy to establish, m the hydraulic mean depth is less so. It is the relationship between the area of water flowing in a drain and its wetted perimeter, shown in Figure 8.54. For proportional depths of flow, Table 8.10 is a convenient reference.

Chezy's coefficient has several derivations relating to friction within the pipe. The metric conversion of a formula attributed to Robert Manning (circa late nineteenth century) provides acceptable values for C. Chezy's coefficient:

$$C = \frac{1}{n} \times m^{1/6}$$

where n = pipe roughness coefficient (0.010 is acceptable for modern drainage)
$\quad m$ = hydraulic mean depth.

Table 8.10 Hydraulic mean depth (m) for proportional depths of flow

Proportional depth	Pipe diameter multiplied by:
Full	0.25
0.75	0.30
0.66	0.29
0.50	0.25
0.33	0.19
0.25	0.15

e.g. A 225 mm drain flowing two-thirds full bore:

$$m = \text{pipe diameter} \times \text{multiplier}$$

$$= 0.225 \times 0.29 = 0.065\ 25$$

Using the previous example of a 300 mm diameter sewer flowing 0.75 proportional depth, at a velocity of 1.526 m/s, the minimum gradient can be established from Chezy's formula:

$$V = C\sqrt{m \times s}$$

Manning's interpretation of C,

$$C = \frac{1}{n} \times m^{1/6} \quad (m = 0.3\ \text{m} \times 0.30 = 0.09\ (\text{see Table 8.10}))$$

$$= \frac{1}{0.01} \times 0.09^{1/6}$$

$$= 100 \times 0.67 = 67$$

Substituting in Chezy's formula,

$$1.526 = 67\sqrt{0.09 \times s}$$

$$\left(\frac{1.526}{67}\right)^2 = 0.09 \times s$$

$$s = 0.0057, \quad \text{where } s = \frac{1}{x}$$

Therefore

$$x = \frac{1}{0.0057} = 175$$

The minimum slope or gradient will be

$$\frac{1}{175} \quad \text{or 1 in 175}$$

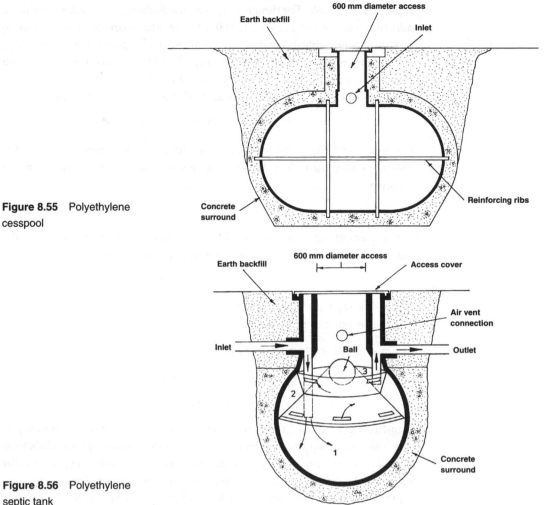

Figure 8.55 Polyethylene cesspool

Figure 8.56 Polyethylene septic tank

Foul water containment and processing

Cesspools, septic tanks and petrol interceptors

Cesspools and septic tanks are acceptable methods for foul water containment and treatment where main sewers are not available. A cesspool is simply an impervious container which is emptied periodically. A septic tank is a small-scale sewage processing plant, which operates principally by the decomposition of solids by anaerobic bacterial activity in the absence of dissolved oxygen.

Both cesspools and septic tanks are produced in reinforced plastic materials for convenient location in prepared excavations. The principles of location with concrete protection and retention are shown in Figures 8.55 and 8.56.

Cesspools

Although cesspools are simple in construction, they have the disadvantage of requiring periodic emptying when full. This is costly, therefore a large capacity is desirable. For design purposes, the Building Regulations require a minimum capacity of 18 m^3 or 18 000 l. At the other extreme, it is unlikely to be structurally viable to build a cesspool over 50 m^3 capacity. Effective storage calculations may be based on the figure of 150 l per person per day, with emptying taken as a minimum of 45-day intervals.

For example, a 5-person household at 45-day emptying intervals,

$$5 \times 150 \times 45 = 33\,750\,\text{l or } 33.75\,\text{m}^3$$

Location should be some distance from habitable accommodation, preferably not closer than 15 m. Vehicular access will need to be within 30 m for emptying.

Septic tank

Periodic emptying is not as critical as for cesspools, but they should be desludged annually. Capacity is based on the simple formula:

$$C = (180P + 2000)$$

where C = capacity (l) (min. 2720 l)

P = number of persons normally using the tank (min. 4).

For example, a five-person household,

$$C = (180 \times 5) + 2000$$

$$= 2900\,\text{l or } 2.9\,\text{m}^3$$

Construction is in three separate compartments. The first compartment separates solids from liquids; heavier solids settle as sludge and lighter solids form a surface scum. The scum is a solid crust which excludes oxygen, thereby encouraging anaerobic action to reduce the volume of solids. The middle compartment is half the volume of the first and completes any further separation of solids that may be necessary. The outlet from the third compartment could feed a small biological filter of broken stone or rubble, where aerobic bacterial action will complete the process before the water is discharged. Discharge may be into a stream or river, or by percolation into underground porous strata by means of inverted perforated land drainage pipes. Whatever is selected will require permission from the local water authority and they will probably examine treated water samples on completion of the installation. Depending on location, direct outflow from the primary treatment unit shown in Figure 8.56 is rarely acceptable as this would represent only about a day's treatment.

Oil and petrol interceptors These chambers are essential where the surface water drains from garage forecourts, industrial premises, etc. and flows into a drain which discharges into a river or other watercourse. The construction may be traditional brick and concrete or reinforced plastic in three compartments shown in principle in Figure 8.57.

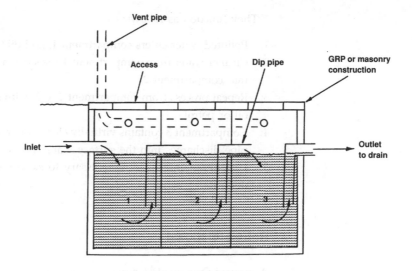

Figure 8.57 Petrol/oil
interceptor, operating
principles

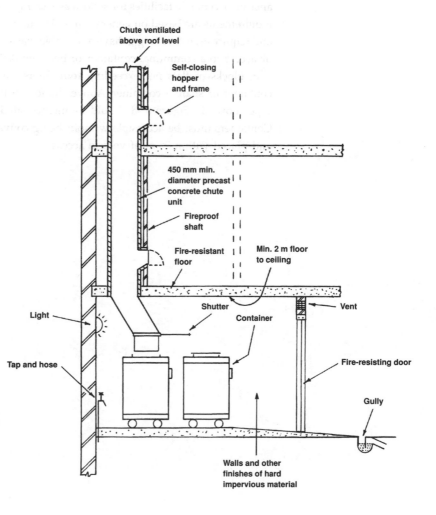

Figure 8.58 Refuse
chute

They function as follows:

1. Polluted water enters compartment 1, and oil/petrol floats to the surface.
2. Cleaner water from compartment 1 rises up the dip pipe and is displaced into compartment 2.
3. Repeat process from compartment 2 to 3, with smaller amounts of surface oil or petrol.
4. Compartment 3 contains virtually clean water and the outlet from this dip pipe discharges into the surface water sewer.
5. Periodic maintenance is necessary to skim the surface of oil and petrol deposits.

Refuse disposal

Approved Document H, section 4 of the Building Regulations, provides guidance on acceptable facilities for solid waste storage. Domestic low-rise housing requirements are based on generation of 0.09 m^3 refuse per dwelling per week and require each dwelling to have a movable waste container of at least 0.12 m^3 or access to a communal container of between 0.75 and 1 m^3 capacity.

In blocks of flats not exceeding four floors, each flat may have its own container or share a container. Above the fourth floor a chute system of the type shown in Figure 8.58 should be incorporated into the building design. Containers must be accessible without being conveyed through the building and placed within 25 m of vehicle access.

9 Security and fire protection

Security

Intruder alarm systems

As the pattern of social misbehaviour and disregard for other people's property by a minority of the population has increased, the interest in intruder alarms for both domestic and commercial buildings has developed considerably. They comprise a bell or siren, a power supply from mains electricity or batteries, switches and a control panel. These components are linked by electrical wiring or a circuit.

Circuits

Alarm system circuits link a number of components as shown in the block diagram, Figure 9.1. The wiring arrangement can be either:

- open circuit, or
- closed circuit.

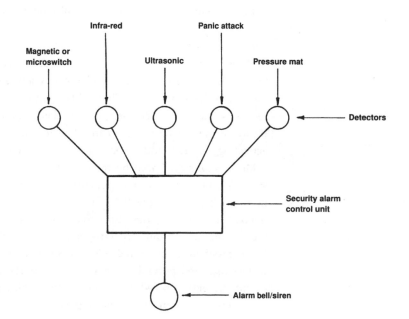

Figure 9.1 Intruder alarm, block diagram

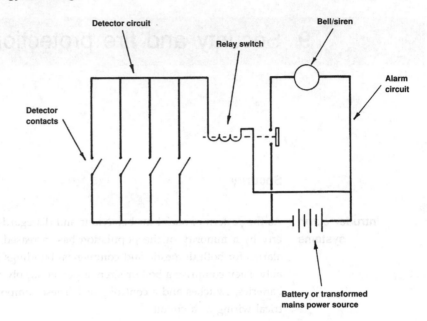

Figure 9.2 Open circuit
intruder alarm

Open circuit

With the open circuit shown in Figure 9.2, the system is activated when a pair of detector contacts wired in parallel are closed. This energises the coil in the alarm relay switch and closes the contacts of the alarm circuit to effect the sounder. Detectors range from simple microswitches to sophisticated heat and vibration sensors, some of which are considered later in this chapter. The open circuit has the disadvantage that it can be rendered inoperative if the detector wiring is deliberately cut or accidentally damaged.

Closed circuit

With the closed circuit, detector contacts are wired in series as shown in Figure 9.3. Each pair of contacts is closed to complete a circuit which energises the coil in the alarm relay. This holds open the contacts until one of the detectors is operated to de-energise the relay coil and complete the bell circuit. If any of the wiring is cut in the detector circuit, it has the same effect as activating a switch and de-energising the relay coil.

Most circuits are based on the closed type and include additional features such as a tamper loop and personal attack button, as featured in the connection diagram (Figure 9.4). The tamper loop is a continuous conductor, linked through every detector contact. Whether contacts are effected or not, the continuously monitored tamper loop will activate the alarm if any system component is interfered with. Personal attack or panic buttons can be a battery-operated portable unit with alarm activated by radio signals.

Exit and entry procedure involves a time delay between setting the system and exiting and entering a building and switching the system off.

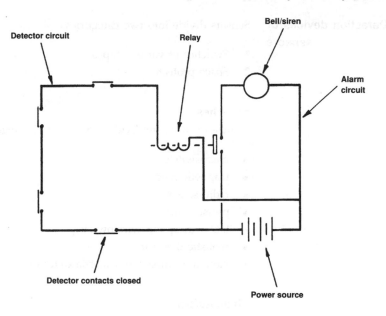

Figure 9.3 Closed circuit intruder alarm

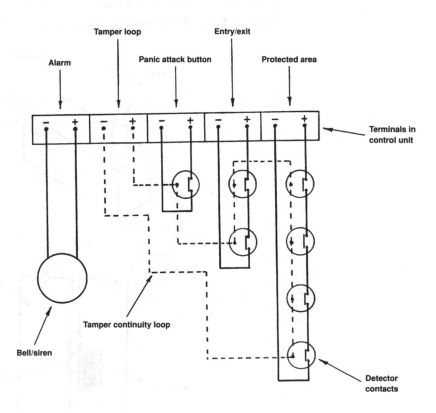

Figure 9.4 Schematic intruder alarm circuit

Detection devices or sensors

Sensors divide into two categories:

1. Switches of various types
2. Space protectors.

Switches

Switches can be subdivided into several categories, namely:

- microswitch
- magnetic reed
- radio sensor
- pressure mat
- taut wiring and window strip
- acoustic detector
- vibration, impact and inertia detectors.

Microswitch

This is the simplest of circuit activation devices and operates by means of a spring-loaded plunger. It is fitted between a door or window and its frame as shown in Figure 9.5. When the plunger is moved, it makes or breaks contact

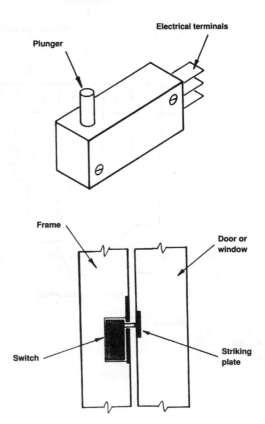

Figure 9.5 Microswitch detector

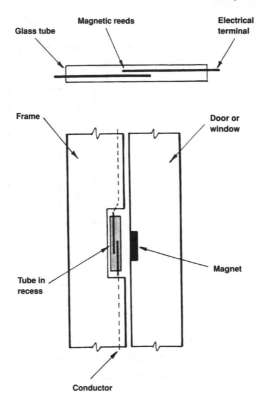

Figure 9.6 Magnetic reed switch

depending on whether an open or closed circuit is used. Perhaps a more familiar application is the device attached to the door recess in a motor car, which operates the interior light with door movement.

Magnetic reed

Microswitches are exposed and can be subject to tampering as well as wear of the moving parts. An improvement that can be used in similar situations is the magnetic reed switch. These consist of a pair of metal strips which overlap and are sealed in a glass tube as shown in Figure 9.6. The metal strips are dissimilarly polarised so that they make contact in a magnetic field. With the switch in a frame, and the magnet in the door or window, opening will separate the magnet from the switch, de-magnetise and separate the contacts to disconnect in a closed circuit.

Radio sensor

These comprise a transmitter and receiver. The transmitter is conventionally switched and can be fitted to doors or windows activating a signal from a tiny battery-powered unit. The control unit, on receiving the signal, activates the alarm circuit. Although the system is limited to the transmitter battery life, it

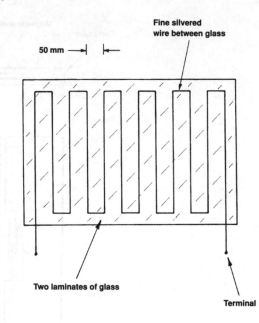

Figure 9.7 Wired glass

does eliminate the need for much 'hard wiring'. Thus, 'free wired' systems have the advantage of temporary applications, considerable flexibility in use and the option for portable personal panic attack buttons.

Pressure mat

These contain a 'sandwich', having two sheets of metal foil separated by a core of perforated foam rubber or plastic. The whole unit occupies an area of about 0.3 m² and is sealed in plastic for insertion under a carpet. Overall thickness can be up to 6 mm, therefore the underlay may need removal to make room for the mat.

Pressure connects the metal foil terminals through the voids in the core, to complete an open circuit system. Strategic location would include: in front of TV and video equipment, under windows and behind doors.

Quality varies considerably with these products, as does the amount of pressure to effect contact. Some can be activated by the household pet, while others require a substantial load over a full shoe sole. Long-term efficiency should be considered, as in time the foil may become damaged and the perforated core could deteriorate.

Taut wiring and window strip

Continuous wire in a closed circuit can be installed around safes, in the floor, wall and ceiling in addition to incorporation within panes of glass as shown in Figure 9.7.

The conductors are usually plastic-coated copper wires, unless installed in a window. Here, the silvered wire is very fine and arranged in rows about

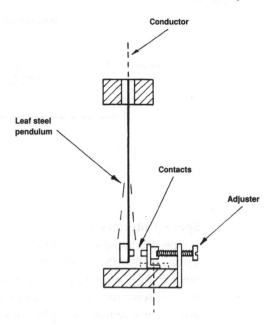

Conductor

Leaf steel
pendulum

Contacts

Adjuster

Figure 9.8 Vibration
detector

50 mm apart between two bonded glass sheets. The effect is similar to that of
a car rear heated window, while still retaining good visibility.

A similar alternative uses self-adhesive lead or aluminium foil, attached
to the vulnerable peripheral areas of glazed windows and doors. Wired as a
closed circuit, fracture of the foil opens an alarm circuit. The obvious visibility
of the foil can be an aesthetic disadvantage, but it does provide an obvious
deterrent to potential intruders.

Acoustic detector

These are most commonly applied to industrial and commercial premises, and
are usually tuned to the sound frequency of breaking glass. Receivers contain
a microphone, an amplifier and output relay that can be adjusted to various
sounds other than breaking glass, e.g. drilling, hammering, voices.

Vibration, impact and inertia detectors

Vibration and impact detectors function when two contacts meet as a result of
compression waves from hammering or impact on the structure. The detector
contains a leaf spring or pendulum which amplifies the movement to make or
break contact, depending on whether an open or closed circuit is used. Figure
9.8 shows an example.

Inertia detectors respond to gentle movements otherwise undetected by
vibration and impact devices. This could be the levering or bending of structural features or the leaning of periphery fences if climbed on. The pivotal
device shown in Figure 9.9 forms part of a closed circuit, where movement of
the weight breaks continuity as the upper contact moves away.

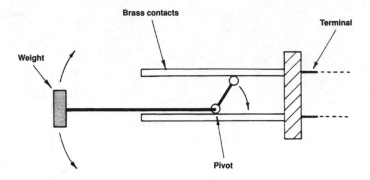

Figure 9.9 Inertia
detector

Space protectors

The preceding detection devices or perimeter defences differ from space protectors, in that they detect intruders during entry or while they are in the process of penetrating property. Space protectors or volumetric sensors as they are sometimes called, deter or keep intruders out by recognising their presence before they have physical contact with a building or its periphery.

The principal space protectors are:

- ultrasonic
- microwave
- active infra-red
- passive infra-red.

Ultrasonic

These detectors use high-frequency sound emitters, generated from an electronic oscillator. The frequency is between 20 and 40 kHz, which is above the upper limit of human hearing of about 16 kHz in a healthy young person. The ultra high-frequency sound is picked up by a receiver containing a microphone with amplifying and processing circuits.

In the absence of movement, the direct and reflected sounds remain at the same frequency, but if the polar range or pattern shown in Figure 9.10 is encroached, the sound frequency will change. Thus, the microphone picks up two frequencies, the original and that reflected from the intruder plus a frequency which is a mixture of the two. This latter 'beat note' is detected as an irregularity and effects an output relay in an alarm circuit. These systems are most effective up to about 3 m and are best arranged with the polar lobe projecting towards an entry gate or door.

These detectors are very sensitive to forward and backward movement, but have little lateral range. They can be susceptible to false interpretations from ultrasonic sounds originating in computer peripherals and other electronic office equipment, in addition to vehicle brakes, compressed air lines and other physical sources.

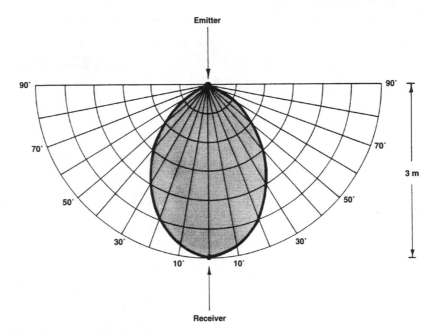

Figure 9.10 Utrasonic detection zone

Microwave

The operating principles of microwave detectors are similar to ultrasonic detectors, except they deploy extremely high radio waves oscillating at a standard 10.7 GHz; 1.48 GHz is also used in small rooms and security compartments. Transmitter and receiver are housed in the same unit; a small plastic box normally mounted at high level to extend waves over large areas such as warehousing and open plan offices. An intruder penetrating the radio microwaves shifts the frequency to activate the alarm.

These devices have the advantage of penetrating 'soft' materials such as wood, plastic, fibre-board, etc., but will reflect off metals. They are unimpeded by draughts, air turbulence and sound waves, therefore false alarms are less likely than with ultrasonics.

Active infra-red detectors

These are sometimes referred to as optical systems. They utilise the infra-red parts of the electromagnetic spectrum that lie just below visible light. Figure 9.11 illustrates the application of the invisible light beam from a transmitter to a receiver containing a photoelectric cell. An intruder crossing or interrupting the beam of light will deactivate the photo cell, to effect the associated alarm circuit. Intruder confusion is ensured by making transmitter and receiver look identical and by fitting dummy units. The smarter intruder may try to beat the system by shining a torch light into a receiver, but this can be overcome by pulsing the light transmission at approximately 200 pulses per second.

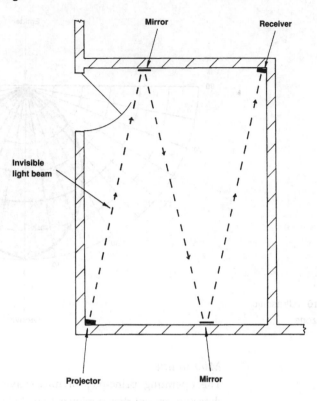

Figure 9.11 Infra-red
optical detector

Beam distances up to 300 m are possible with reflection off mirrors to provide comprehensive coverage of corridors and rooms, but each reflection will reduce efficiency and range by about 25 per cent. External use is limited by the effects of dense fog, airborne objects and birds crossing the beam.

Passive infra-red detectors (PIR)

The PIR detector is a relatively new security device, enjoying an enormous growth in market share over the past 15 years. It works on the principle of highly sensitive ceramic detectors receiving infra-red radiation from a moving body. The radiation focuses through vertical facets on a curved plastic lens on to two sensors as shown in Figure 9.12. The alteration of image from one sensor to another, generates an electrical differential which activates an alarm or high wattage illumination.

The facets produce zones of detection over distances up to about 20 m, with some narrow variations for corridors up to 40 m long. Some typical examples of patterns and ranges are illustrated in Figure 9.13.

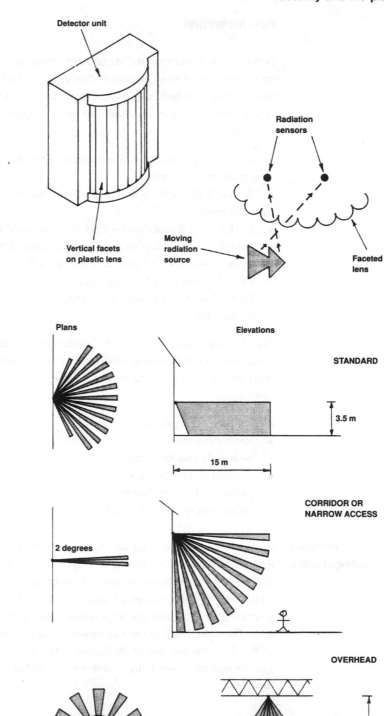

Figure 9.12 Passive infra-red detector.
Note: Some lenses are smooth, with faceted reflectors inside the unit

Detector unit

Radiation sensors

Vertical facets on plastic lens

Moving radiation source

Faceted lens

Plans

Elevations

STANDARD

3.5 m

15 m

CORRIDOR OR NARROW ACCESS

2 degrees

OVERHEAD

4 m

Figure 9.13 PIR beam patterns

Fire protection

Protection of a building and its contents from damage by fire can be divided into passive and active categories. Passive control refers to the means by which the design of the building, its structure, fabric, components and their installation, resist fire. It is an extensive and complex subject incorporating the influence of:

- the building insurers, whose requirements may extend beyond legislative minimum standards
- the local fire officer, as for the insurers plus concern for fire-fighting accessibility
- the Building Regulations – Part B, mandatory requirements for purpose grouping of buildings and compartmentalisation within buildings, as well as fire resistance and potential fire spread of materials
- BS 5588, planning of escape routes
- the local authority planning department and the Health and Safety Executive standards.

The aforementioned is largely the domain of construction technology, which is not the brief of this book. However, active means of fire protection and fire engineering is a specialised area of building services and the following will be considered:

- portable extinguishers
- alarm detection
- hose reels and hydrants
- automatic extinguishers
- pressurised escape routes
- smoke extraction and ventilation.

Portable extinguishers

Buckets of water and sand are no longer accepted as adequate provision for first aid fire control. Colour-coded cylinders containing compressed liquids and gases appropriate to various sources of fire, are now standard fire-fighting equipment in all commercial and public buildings. Personnel should be instructed to recognise the applications and Table 9.1 provides some guidance. The objective is to remove or sufficiently reduce at least one component of the fire triangle shown in Figure 9.14. A correctly used portable extinguisher will cool a small fire to remove the heat or smother it to prevent access for oxygen.

Alarm detection

Fire alarm circuits may be of the open or closed type, previously described for intruder alarms (see Figures 9.2 and 9.3). The exception is that the contacts are usually in wall-mounted, break-glass switch units. Most installations will include an indicator board to locate the source of alarm, and

Table 9.1 Portable fire extinguishers

Application	Colour-coded cylinders and content				
	Red	Red and Cream	Red and Black	Red and Blue	Green
	Water	Foam	Carbon dioxide	Powder	Vapourising liquid
Paper, wood, and textiles	✓	✓	✓	✓	✓
Flammable liquids, gases, paints and greases		✓	✓	✓	✓
Electrical hazards			✓	✓	✓
Machinery, plant and vehicles			✓	✓	✓

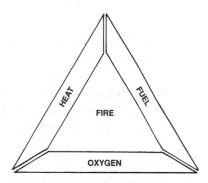

Figure 9.14 Fire triangle

Figure 9.15 shows the circuitry principle with a battery power source or transformed mains supply of 24–60 V DC. Alarm location should be in common access positions and no person should have to travel more than 30 m to raise an alarm. Landings, lobbies and corridors are the best location, with clearly defined, red-painted call buttons 1.5 m above floor level.

Automatic fire detectors

Automatic fire detectors are necessary to indicate location of the outbreak of a fire, to operate alarm bells and to communicate with the local fire authority. They are varied in operating characteristics, but may have:

- a bimetallic strip
- an ionisation chamber

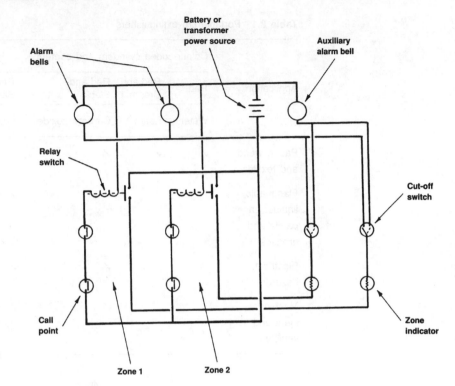

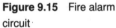

Figure 9.15 Fire alarm circuit

- light-scattering devices, or
- a laser beam.

Bimetallic strip
This is the simplest and contains a strip which responds to temperature increases. It deforms to bend across two electrical contacts to complete a circuit.

Ionisation chamber
This more sophisticated device ionises air by radiation, to encourage a small electric current across two electrodes. When smoke enters the chamber it reduces the current, and this irregularity is sufficient to effect an alarm relay.

Light-scattering devices
Figure 9.16 shows the operation of this small wall- or ceiling-mounted unit. Under normal conditions it has a light source projecting its beam into a light trap. When smoke enters the unit, the light is scattered by reflection off the smoke to fall on a photoelectric cell which energises an alarm relay.

Laser beam
These are an economic solution to provision of fire detection in large areas, as the concentrated beam can be effective over distances up to 100 m. Light beams can be visible or infra-red, and target on an opposing photoelectric cell

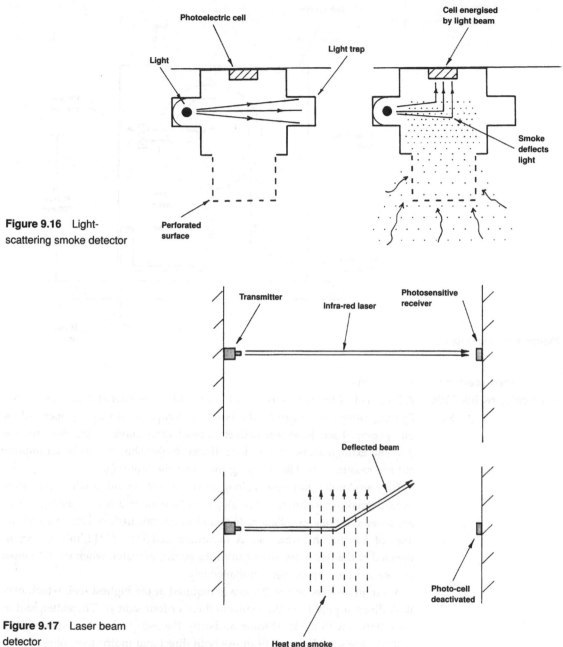

Figure 9.16 Light-scattering smoke detector

Figure 9.17 Laser beam detector

as shown in Figure 9.17. Smoke obscuration or air turbulence caused by heat, deflects the beam to de-energise the receiving cell which activates an alarm relay. They may also double as intruder alarms.

Note: Location of detectors should not exceed one per 100 m² of floor area, but the building insurers and the fire service may require them closer.

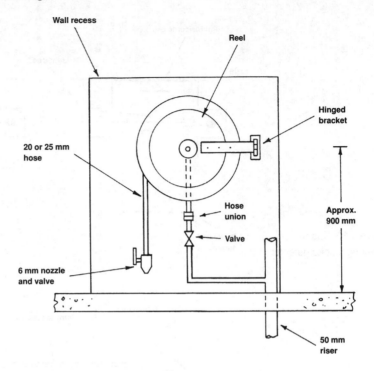

Wall recess

Reel

Hinged
bracket

20 or 25 mm
hose

Hose
union

Approx.
900 mm

Valve

6 mm nozzle
and valve

50 mm
riser

Figure 9.18   Hose reel

**Hose reels and
hydrants, re BS 5306
Pt 1**

Hose reels

A hose reel of the type shown in Figure 9.18 is considered a first aid to fire fighting, intended for use by the building occupants. It may be operated by one person. Each hose reel delivers considerably more water than several portable extinguishers, and with continued replenishment can be an important fire resource capable of saving lives and the building.

They are located in recesses along corridors and provided with up to 45 m of reinforced rubber hose, so that all parts of a floor area not exceeding 800 m^2 are covered by one installation. Included in the calculations can be an allowance of 6 m for the water jet. A minimum delivery of 24 l/min is recommended at the reel most distant from the source of water, when the two most remote reels are operating simultaneously.

A minimum pressure of 200 kPa is required at the highest reel, which may limit direct supply from the mains to three or four storeys. Thereafter, and in consultation with the local water authority, the use of a break or suction tank will be necessary. Figure 9.19 shows both direct and indirect supplies, with a minimum capacity of 1125 l in the break tank. Inlet supply float valves are at least 50 mm nominal bore and duplicate pumps (stand-by and duty) are operated by pressure switch and pressure drop in the system when a hose reel valve is opened.

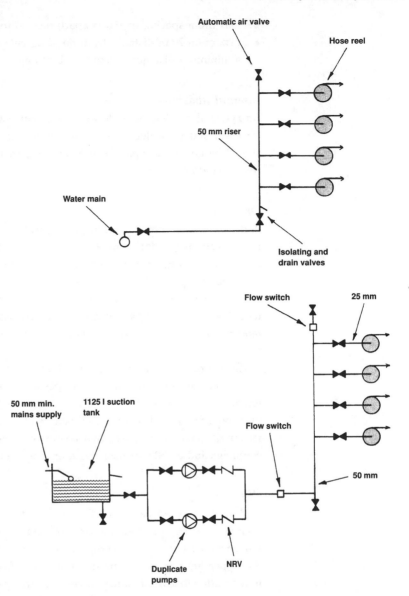

Figure 9.19 Hose reel installation

Hydrants

External hydrants

Fire hydrants for general use are installed on the underground water mains serving towns, villages, etc. For new building projects, insurance companies and local fire authorities may require an independent private hydrant system, particularly if the building's fire risk is high.

The hydrant valves should attach to a ring system of supply, with more than one source from the water authority's main. Other recommendations include:

- maximum spacing of 150 m apart, next to roads
- maximum 70 m distance from building entry
- a minimum distance of 6 m to a building.

Internal hydrants

As a general installation guide, one rising main (wet or dry) must be provided for every 900 m² of floor area. Where more than one riser is required, they should be no further apart than 60 m. No part of a floor must exceed 60 m from a landing valve.

Dry riser

A dry riser is an empty vertical pipe complete with hydrant outlets at each floor. A coupling within 18 m of an access road is provided at ground level for fire fighters to connect their hose. In effect, it is an extension of the fire fighters' hose and is appropriate in unheated buildings and structures between 18 and 60 m in height, where prompt attendance from the fire authority is guaranteed. Above 60 m, wet risers must be installed because of the limitations of fire tender pumps and the need for water to be immediately available at hydrant valves.

The dry riser installation shown in Figure 9.20 has a 100 mm nominal bore galvanised steel pipe for one hydrant per floor and a 150 mm pipe if two outlets are fitted. At ground level the inlet is provided behind a red painted wired glass box, fitted flush with the outside wall and clearly labelled DRY RISER INLET. A 100 mm riser has two 65 mm breeches with instantaneous hose couplings and a 150 mm riser has four. The riser is electrically earthed.

Wet riser

The wet riser is permanently filled with water to supply hydrant valves on each floor. A 100 mm riser can serve one outlet per floor and a 150 mm pipe will be sufficient for two. Pressure requirements are between 400 and 500 kPa, the upper limit to protect fire-fighting hoses from rupturing. Orifice plates may be fitted to lower landing valves to restrict pressure. Direct supplies from the street main must also provide a flow rate of 25 l/s. If this is unrealistic, a break or suction tank of 45 000 l capacity must be installed from which duplicate pumps (one electric, one diesel or two electric and a stand-by diesel) provide the minimum delivery. Figure 9.21 shows the application for each 60 m height in a building. Beyond this additional storage and further pumping will be necessary.

The wet riser operates as follows. The duty pump is activated by a fall in water level registering at the pipeline switch. The flow and pressure switch responds to water flowing and when all hydrant landing valves are closed, lack of hydraulic movement engages the flow and pressure switch to disconnect the pump.

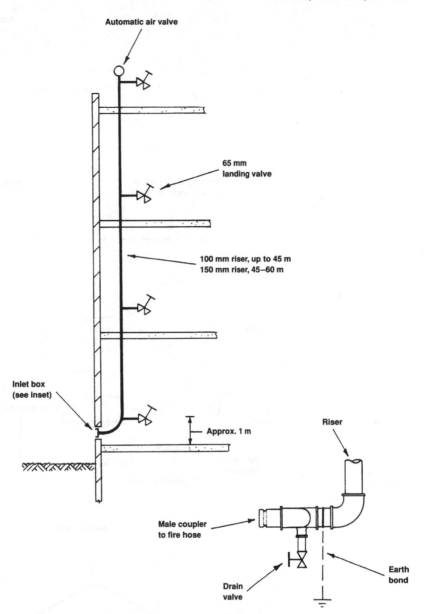

Figure 9.20 Dry riser

Foam installations

Foam extinguishing systems are preferred for application to ground floor and basement boiler plant rooms and fuel storage areas, where oil is the heating medium. Foam is generated by special mixing equipment for fire fighters to connect through an external foam inlet box found about 500 mm above ground level. It is similar to that described for a dry riser, but clearly labelled FOAM INLET.

The system shown in Figure 9.22 has one inlet into a 63 or 75 mm nominal

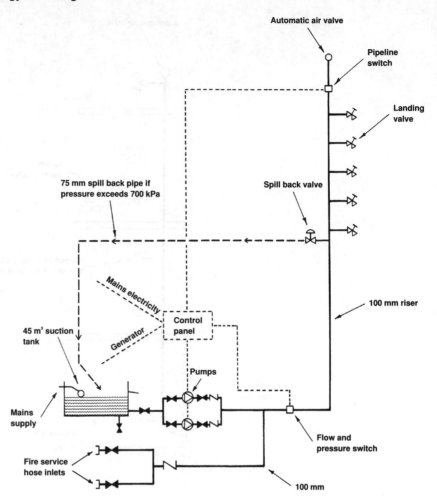

Figure 9.21 Wet riser

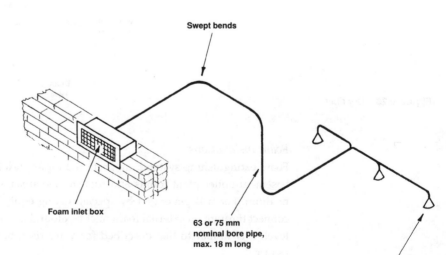

Figure 9.22 Foam inlet
and pipework

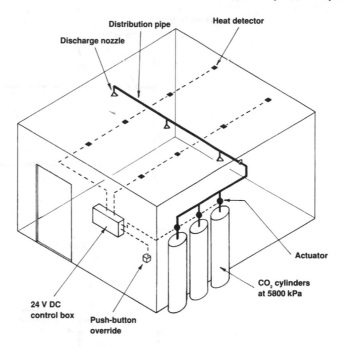

Distribution pipe

Heat detector

Discharge nozzle

Actuator

CO_2 cylinders
at 5800 kPa

24 V DC
control box

Push-button
override

Figure 9.23 Installation
of carbon dioxide

bore pipe, not exceeding 18 m in length containing a maximum of three outlets. The maximum aggregate cross sectional area of outlets is 3200 mm².
Terminals may be about 1 m above oil-fired boiler burners and 150 mm above
leakage level of stored oil.

Automatic
extinguishers

Gas extinguishing systems

Pressurised halon and carbon dioxide gas is stored in cylinders. It is activated
by smoke or heat detectors to effectively smother fires and substitute where
water may be considered too damaging. They are colourless, odourless gases,
halon being more effective with five times the density of air, while carbon
dioxide has only 1.5 times air density. However, halon or bromochlorodi-
fluoromethane, abbreviated BCF, has lost favour in recent years as it has a
depleting effect on the ozone layer.

Carbon dioxide has been used as an extinguishing agent for many years,
notably in portable extinguishers. Its non-conductive properties are ideal for
electrical hazards as well as general applications to gaseous, petroleum, oil
and carbonaceous fuels. Being denser than air, it effectively reduces the oxygen content of air from a normal 21 per cent to about 15 per cent, efficiently
disabling the combustion triangle.

Figure 9.23 shows a total flooding system, where heat or smoke detectors
effect the gas discharge through actuators fitted to the storage cylinders. Total
flooding relies on gas containment by peripheral means, but where fire hazards are isolated, a local application system will discharge gas directly over
burning material. The actuator or automatic valve fitted to each cylinder is
operated directly by the gas pressure, but this can be bypassed manually.

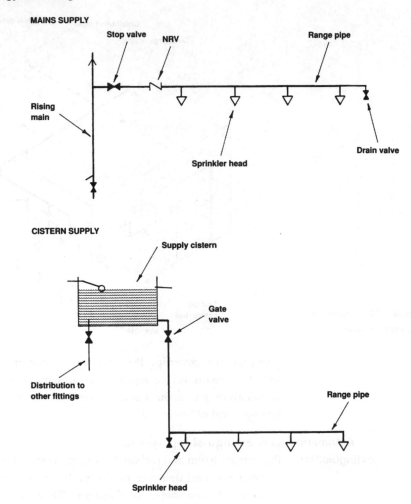

Figure 9.24 Simplified sprinkler installation

Water sprinklers

An automatic water sprinkler system to control the outbreak of fire was first developed by an American, Henry Parmalee, during the latter part of the nineteenth century. During this time, further developments and improvements were conducted by another American, Frederick Grinnell, and it is his name that continues to be associated with glass bulb sprinkler heads. The simplest of applications is shown in Figure 9.24, with sprinkler heads attached to a rising main or a cistern supply.

Sprinkler heads may be:

- glass bulb
- fusible link
- chemical, or
- open.

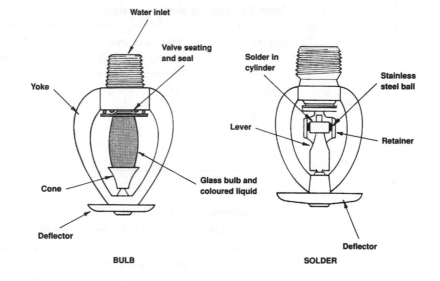

Figure 9.25 Sprinkler heads

Table 9.2 Sprinkler bulb colour and rupturing temperature

Bulb liquid colour	Rupturing temperature (°C)
Orange	57
Red	68
Yellow	79
Green	93
Blue	141
Mauve	182
Black	204/260

The most frequently specified glass bulb and fusible link heads are shown in Figure 9.25. The bulb type has its valve firmly seated by a sealed glass tube containing volatile fluid, which expands in the presence of heat to burst the tube and release water from the associated pipework. Fluid operating temperatures correspond to a colour coding listed in Table 9.2.

The fusible link type has levers retained in position by solder which can be specified to various melting temperatures. On heating, the solder melts and the levers spring apart to release the retained water. A small block of chemical material which melts at a predetermined temperature is a variation on the soldered principle.

Open sprinkler heads are water spray devices which do not operate in isolation. They are part of a bank of heads attached to a dry pipe which charges with water when a smoke or fire detector activates a motorised mains water valve. This is otherwise known as a deluge system and is applied where potential for intense fires having a rapid rate of propagation is possible.

Table 9.3 Spacing of sprinkler heads

Hazard	Max. floor area per sprinkler (m²)	Max. distance between sprinklers (m)
Light	21	4.6
Ordinary	12	4.0 (standard spacing)
		4.6 (staggered, see Figure 9.35)
High	9	3.7

The systems shown in Figure 9.24 are largely notional and lack the sophistication and control acceptable to the influencing authorities. In the UK, the BSI document BS 5306 Pt 2 and the Loss Prevention Council's Technical Bulletins now combine to incorporate the previously accepted rulings of the Fire Offices Committee (FOC). In 1969 the FOC established that the occupancy of a building rather than only its construction should be the major consideration for sprinkler system design.

Design considerations
The type of business and procedures undertaken in a building will determine the fire risk category. These are classified as light, ordinary and high hazard. Examples of light hazard include non-industrial occupancies such as hospitals, hotels, institutions, colleges, museums, etc., ordinary hazard would be production engineering, breweries, broadcasting studios, restaurants, etc. and high hazards are fireworks factories, paint and plastics manufacture and other volatile chemical and fluid-operating premises.

Subdivisions occur within these three overall categories and numerous design tables are generated to provide information on water flow rates, pipe sizes, disposition of sprinkler heads and pipework configurations. Reference to BS 5306 Pt 2 'Specification for sprinkler installations' is recommended for application to particular situations. As a basis for design, Table 9.3 provides the spacing and area allocation of sprinklers for the three hazard groupings.

Sprinkler systems
There are a variety of systems, each associated with different building functions and variations in occupancy:

- wet
- dry
- alternate wet and dry
- tail end
- pre-action
- recycling.

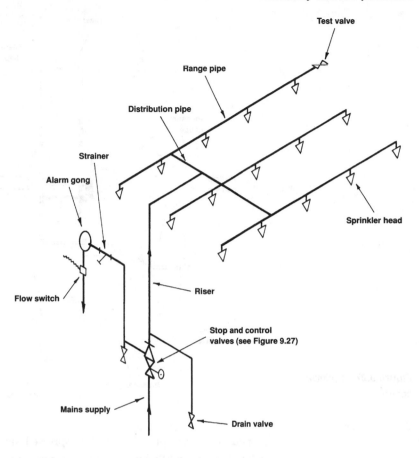

Figure 9.26 Wet sprinkler installation

Wet system

This, the most popular of systems, has all pipework permanently charged with water, for an instant response when a sprinkler head fractures. It is suited to building interiors that remain above freezing and do not exceed 70 °C. Figure 9.26 shows a schematic layout of distribution and range pipes with isolation and alarm valves. Figure 9.27 details the controls which include a special clack or clapper valve. This rises as water flows to a fractured sprinkler head, simultaneously releasing a small quantity of high-velocity water into a small pipe to physically rotate the alarm gong and to activate an electrical alarm flow switch. Small orifices in the valve clapper allow variations in mains pressure without activation of the alarm. The maximum number of sprinklers on one set of control valves is 1000.

Dry system

Dry pipe systems are used in unheated buildings such as warehousing and in sub-zero temperature production environments. They are also used in bakehouses and similar high-temperature situations where the ambient temperature may exceed 70 °C. It differs from the wet system in that the alarm valve

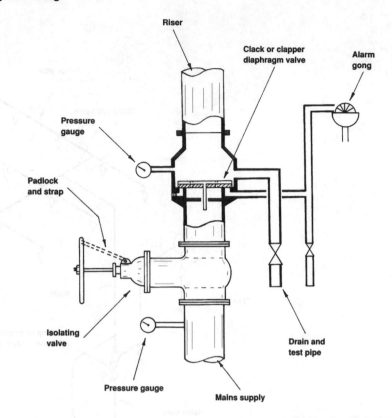

Riser

Clack or clapper
diaphragm valve

Alarm
gong

Pressure
gauge

Padlock
and strap

Isolating
valve

Drain and
test pipe

Pressure gauge

Mains supply

Figure 9.27 Sprinkler
controls

is modified to accept a charge of compressed air at about 200 kPa above the
valve clapper. Slightly lower mains water pressure is retained below. Sprinkler
heads are installed upright, with range pipes falling slightly to a drain valve,
so that water cannot remain trapped in the pipework after testing. Efficiency
is reduced as the air must be discharged before water can reach a fractured
sprinkler, therefore the maximum number of sprinklers is only 250 in one
system, doubling if a pump is added to rapidly exhaust the air.

Alternate wet and dry system
The quicker response of the wet system is always preferred, but is not always
viable in cold buildings. A compromise is to alternate seasonally, with the
system charged with water most of the year and compressed air in the remaining winter months. Sprinkler heads are upright and the number is limited to that specified for a dry system.

Tail end system
This is a variation on the wet system and the alternate wet and dry system.
As the name suggests, it is appropriate where only part of a building is subject
to frost or extremely high ambient temperatures. An air valve is fitted to one
section or 'leg' of a wet system to maintain a differential between water and

compressed air. As the system is predominantly wet, the maximum number of sprinklers remains at 1000, but no more than 100 may be fitted after the tail end air valve. If there are several tail ends in the system, the overall total of tail end sprinklers must not exceed 250.

Pre-action system

This is principally a dry system working in conjunction with smoke or heat detectors which actuate the control valve. This system acknowledges the water damage potential when a sprinkler is accidentally fractured, not unusual from high-stacked fork-lift trucks in warehouses and other storage situations. If this occurs, only the compressed air discharges to initiate a subsidiary alarm. The main alarm and water flow are dependent on the fire detectors. The maximum number of sprinklers is limited to 500 for light hazard occupancy and 1000 for ordinary and high hazards.

Recycling system

This is a development of the pre-action system. During operation it uses ceiling-mounted heat detectors to close the control valve when the air temperature in the vicinity of a fire falls, due to the cooling effect of the sprinkler. The detectors are rated a few degrees lower than the sprinkler head and when heat is no longer sensed, this is relayed through a 5 min delay before the control valve is closed. A temperature rise reactivates the valve.

This system is used where it is considered appropriate to limit damage from water after a fire has been extinguished, or when a sprinkler is accidentally fractured. A maximum number of 1000 sprinklers are permitted with this installation.

Water supplies

An adequate supply at sufficient pressure is usually obtained from the town water main. Other possibilities shown in Figures 9.28–9.30 include elevated reservoirs or tanks and automatic pumping from a mains-filled break or suction tank. Capacities, pressures and delivery rates vary considerably for hazard groupings, building size, numbers and spacing of sprinklers and pipework configuration. Tables to assist design for specific situations are provided in BS 5306 Pt 2.

Pipework configuration

The building plan shape, disposition of rooms and facilities for pipework accommodation will considerably affect sprinkler arrangements. The optimum is a central riser with uniform lengths of distribution and range pipes, but in practice this is difficult to achieve. Figures 9.31–9.34 show a number of distribution possibilities and Figure 9.35 indicates the spacing of sprinkler heads relative to adjacent walls.

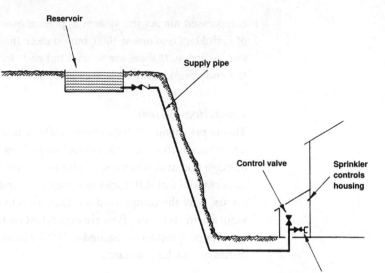

Figure 9.28 Sprinkler system water supply – elevated reservoir

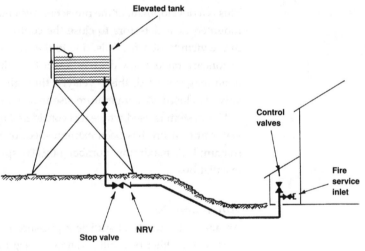

Figure 9.29 Sprinkler system water supply – gravity tank

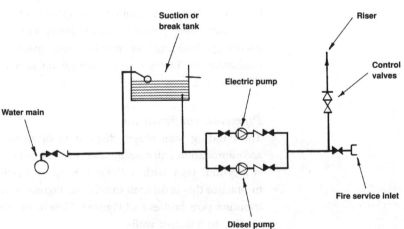

Figure 9.30 Sprinkler system water supply – suction tank and pump

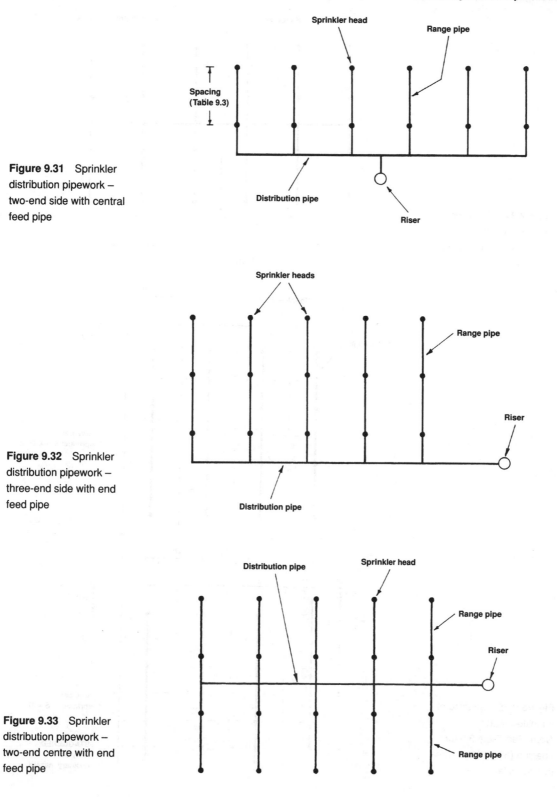

Figure 9.31 Sprinkler distribution pipework – two-end side with central feed pipe

Figure 9.32 Sprinkler distribution pipework – three-end side with end feed pipe

Figure 9.33 Sprinkler distribution pipework – two-end centre with end feed pipe

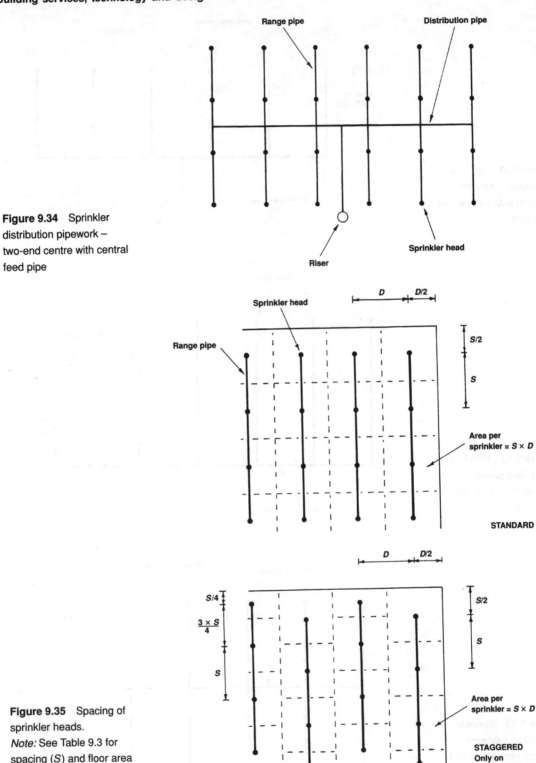

Figure 9.34 Sprinkler distribution pipework – two-end centre with central feed pipe

Figure 9.35 Spacing of sprinkler heads.
Note: See Table 9.3 for spacing (*S*) and floor area per sprinkler

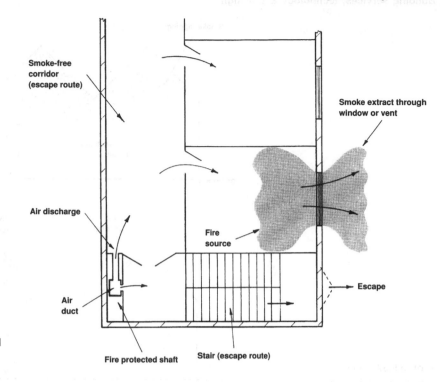

Smoke-free
corridor
(escape route)

Smoke extract through
window or vent

Air discharge

Fire
source

Air
duct

Escape

Figure 9.36 Pressurised
escape route

Fire protected shaft

Stair (escape route)

Pressurised escape routes

The objective is to create greater air pressure in escape routes such as corridors and stairs, than the remainder of rooms in an office block and other large and high-rise compartmentalised buildings. The effect is to contain smoke and fire at its point of outbreak, allowing it to leak out through windows or purpose-made ventilation grills as shown in Figure 9.36.

The normal pressure within a building from convected warm air (stack effect) and wind differentials across a building will have to be overcome. In event of a fire, this too will add further to the internal pressure, the more so the taller the building. Consequently, air pressurisation of escape routes must be at least 25 Pa, possibly as much as 60 Pa in large buildings, but insufficient to impede human progress in an emergency.

If a fire occurs, the detector or alarm will automatically close down all ventilation and air-conditioning plant. Simultaneously it will engage the escape route air pressurisation fan to deliver sufficient volume through fire-protected ductwork. In some buildings the pressurisation plant runs continuously as part of the normal ventilation system, increasing in capacity in response to a fire.

Smoke extraction and ventilation

Automatic smoke ventilation systems originated in factory and warehouses to relieve the build-up of smoke and heat in the event of a fire. With the development of large-area shopping malls, the principle is now applied on a wider scale.

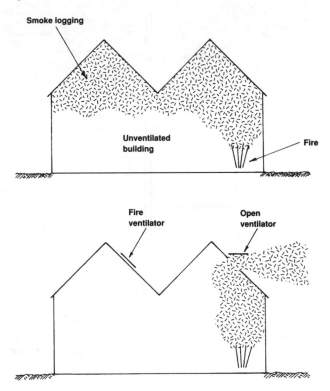

Figure 9.37 Smoke logging and ventilation

The objective is to aid fire control by eliminating smoke, heat, toxic and inflammable gases from the source of a fire, and to retain visibility for escapers and to provide clear access for fire fighters. Figure 9.37 shows the effect of smoke logging in an unvented situation, compared with the visibility available when fire vents operate. Vents function daily as normal roof lights with manual or automatic controls, but if closed during a fire, the heat will melt a spring-loaded fusible link normally set at 70 °C to open the unit. Smoke detectors operating through a relay to release the vent catch are preferable, as very smoky fires may not generate enough heat to fuse a link. The introduction of air nominally enhances the fire, but this is justified by the advantages gained by releasing the smoke, fumes, etc.

Number and size of vents

Table 9.4 can be used to estimate the area of roof vents while providing a smoke-free layer some 3 m above floor level. For example, Figure 9.38 shows a building with floor area of 2000 m² and fire hazard perimeter of 50 m. From Table 9.4, the vent factor is 0.37 m.

$$\text{Vent area} = 50 \text{ m} \times 0.37 \text{ m} = 18.5 \text{ m}^2$$

or

$$\frac{18.5}{2000} = 0.0093 \text{ or approximately 1 per cent of the floor area}$$

Table 9.4 Roof vent, ventilation factors

Floor to centre line of vent height (m)	Ventilation factor (m)
4.5	0.61
6.0	0.43
7.5	0.37
9.0	0.31
10.5	0.27
12.0	0.25
13.5	0.23
15.0	0.21

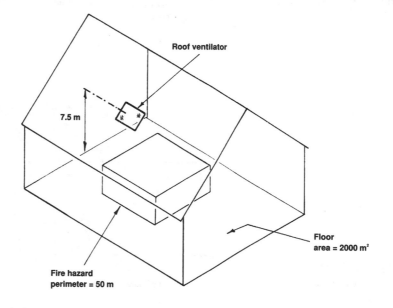

Figure 9.38 Fire vent area calculation

Shop extract and smoke control

Part B of the Building Regulations requires fire prevention systems in shopping centres to be under 'unified ownership and continuing control'. In practice this means that smoke can be controlled by two methods:

1. direct extraction from individual shop units (shop extract), or
2. a common extract system (mall extract).

Shop extract

Responsibility for smoke and fire control can be delegated from landlord to tenant, which is fine if the tenant is fully aware of the undertaking to maintain and test a smoke control system regularly. The extract duct and fan system shown in Figure 9.39 must be fire specified and be separate from any other means of shop ventilation. It must have smoke detection equipment connected

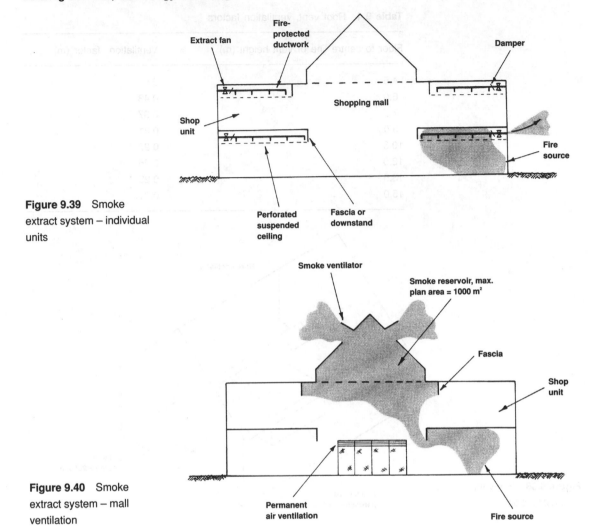

Figure 9.39 Smoke extract system – individual units

Figure 9.40 Smoke extract system – mall ventilation

to a control system to operate dampers and fan. It must also have the facility to shut down any other ventilation system serving the affected shop.

Mall extract

A simpler and much more economic arrangement is a common smoke collection and containment reservoir at the height of a mall. Strategically dispersed smoke detectors can be deployed to effect individual vents in smoke collection zones. Provided the building design accommodates the features shown in Figure 9.40, i.e. restricted smoke reservoir areas and permanent replacement air ventilation, shoppers would have no difficulty escaping unhindered. Fire fighters, too, will have clear access to the source of fire.

If sprinklers are also required, they should be considered at a higher temperature rating than normal (normal is red bulb – 68 °C) to delay the possibility of smoke being cooled and collecting at floor level.

10 Transportation: lifts, escalators and travelators

Vertical transportation of goods and people has its origins in the depths of history, where simple rope and pulley block systems were used. In relation to modern practice, it was the development of steel and the subsequent evolution of high-rise buildings in the latter part of the nineteenth century that promoted the need for assisted movement of people. Concurrently, electricity generation, distribution and application became available and these combinations encouraged Elisha Otis to successfully pioneer vertical transportation systems that still bear his name.

Mechanical transport

The second part of the twentieth century has seen an enormous increase in the number of high-rise buildings, with a corresponding demand for mechanical transportation of goods and people within and around them. The spatial needs and impact of transportation systems have revolutionised building design and such equipment can only be regarded as an integral part of the normal building services process. Accordingly, requirements must be anticipated at an early stage of building design, with full regard to the dependence on other services such as electricity, fire protection, means of escape, co-ordination of installations and longer-term maintenance of the facility.

Movement systems

The following forms of mechanical transportation may be found within, around and in general association with modern buildings and developments:

- lifts
- escalators
- travelators or moving pavements.

Lifts   Lifts are considered a requirement in all buildings over three storeys, less if access for the disabled is in the design brief. A minimum standard of service

Table 10.1 Lift car speeds

Type of lift	Car speed (m/s)
Passenger, up to 4 floors	0.3–0.8
4–9 floors	0.8–2.0
9–15 floors	2.0–5.0
over 15 floors	5.0–7.0
paternoster	Up to 0.4
Goods, to any height	0.2–1.0
Hydraulic, passenger or goods, max. 4 or 5 floors	0.1–1.0

is considered to be one lift for every four storeys with a maximum horizontal distance of 45 m to the lift lobby. Floor space estimates and car capacity can be based on an area of 0.2 m^2 per person.

Lift speeds vary as indicated in Table 10.1, from as little as 0.1 m/s for goods and 0.3 m/s for passengers in low-rise buildings. An upper speed limit for very tall buildings is about 7 m/s due to the limitations of the human ear to absorb atmospheric pressure changes.

Location

Easy means of access for all building users is essential, and this usually means positioning lifts in the central entrance lobby of offices, hotels, flats and the like. In department stores the entrance area could become congested with shoppers waiting for the lifts. As this also represents useful display space it could be advantageous to dispose lifts at opposite ends or corners of the building, with escalators central. Whatever the chosen location, grouping of lifts is essential for user convenience. Figure 10.1 shows some possibilities, but it should be noted that although a small number of large-capacity lifts may be cheaper to install, for operating efficiency several small-capacity lifts are preferred. For example, four 12-person lifts equate to three 16-person lifts in carrying potential, but waiting time could be twice as long due to the number of stops (see round trip time calculations). Average waiting time can be calculated by dividing the round trip time by the number of cars in a bank or group of lifts.

Lift efficiency

An acceptable method for estimating and comparing efficiency and effectiveness of lift installations is obtained by calculating the round trip time (RTT). This is an average period of time for one lift car to circulate, incorporating statistical data for time lost due to stops. It is measured from the time the lift doors begin to open at the main terminal to the time they reopen when the

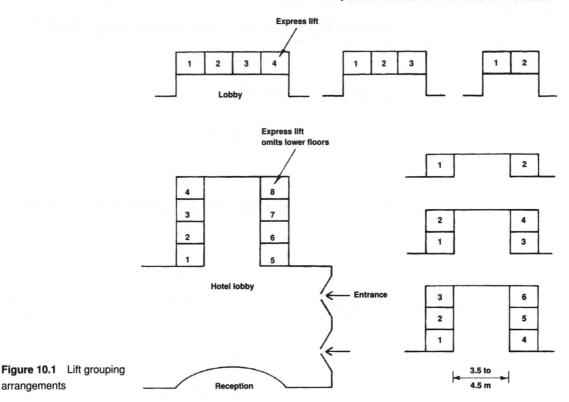

Figure 10.1 Lift grouping arrangements

car completes its cycle. The complete process is a summation of a number of simple calculations relating the lift's function through its cycle, as shown in the following example.

A building having five floors at 3 m floor to floor spacing, a car capacity of six persons and 2 m/s speed of travel:

1. Probable number of stops (S_1):

$$S_1 = S - S\left(\frac{S-1}{S}\right)^n$$

where S = maximum number of stops and n = number of people or car capacity.

$$S_1 = 4 - 4\left(\frac{4-1}{4}\right)^6 = 3.3, \quad \text{i.e. 3 stops}$$

2. Upward journey time (T_u):

$$T_u = S_1\left(\frac{L}{SV} + 2V\right)$$

where L = lift travel, 4×3 m = 12 m, and V = car speed, 2 m/s.

$$T_u = 3\left(\frac{12}{4 \times 2} + [2 \times 2]\right) = 16.5 \text{ s}$$

3. Downward journey time (T_d):

$$T_d = \frac{L}{V} + 2V$$

$$= \frac{12}{2} + [2 \times 2] = 10 \text{ s}$$

4. Passenger transfer time (T_p). Allow 2–3 s per person to transfer, depending on the depth of car. At 2 s:

$$T_p = 2n$$

$$= 2 \times 6 = 12 \text{ s}$$

5. Door opening time (T_o). Assume door speed (V_d) = 0.5 m/s and door width (W) = 1.2 m:

$$T_o = 2(S_1 + 1)\frac{W}{V_d}$$

$$= 2(3 + 1)\frac{1.2}{0.5} = 19.2 \text{ s}$$

6. Round trip time (RTT):

$$\text{RTT} = T_u + T_d + T_p + T_o$$

$$= 16.5 + 10 + 12 + 19.2 = 57.7 \text{ s}$$

Notes:

1. For lift and system comparison, these calculations assume the lift deposits passengers in one direction only and then returns. For a broader perspective, allowance can be made for downward stops, passenger transfer and door operating times.
2. If a group of lifts is being considered, the RTT is calculated by dividing by the number in the group.

Electric lifts

Motor

Lift motors have been powered from a direct current (DC) generator with variable output, because of their greater accuracy in use, i.e. floor levelling and smoother ride, than alternating current (AC) variations. Developments during the 1980s provided a gradual transfer to static converters as replacements for DC generators. Also sufficient improvements in the control of

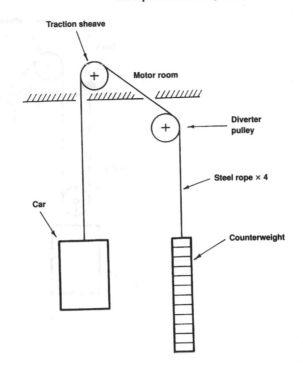

Figure 10.2
Single wrap 1 : 1

AC motors have produced acceptable ride controls and wider use of this alternative. Some installations combine both AC and DC, the traction sheave powered by AC, with brakes and reduction gearing by DC. DC motors are still preferred in these situations as they are more durable in resisting variable demands and frequent stopping and starting.

Roping

High-tensile steel ropes are driven through traction sheaves attached to the motor shaft, a system of pulleys and a counterweight. Various combinations are possible to suit different occupancy requirements. Single wrap arrangements shown in Figures 10.2–10.4 are the simplest, but will be prone to slippage if subjected to heavy loads. Resistance to slippage can be improved by increasing the number of pulleys and the roping ratios, but as the ratio increases, the car speed decreases proportionally as shown in Table 10.2.

To maintain high speeds and sufficient traction, double wrap sheaving shown in Figure 10.5 is necessary. This is achieved by passing the rope from the car over the traction sheave, around the diverter pulley and back over the traction sheave before connecting to the counterweight.

In very tall buildings the effect of bounce and spring from the rope load can be balanced or compensated with ropes suspended below car and counterweight as shown in Figure 10.6. The compensating rope passes around a large sheave which will require special consideration for accommodation in the pit.

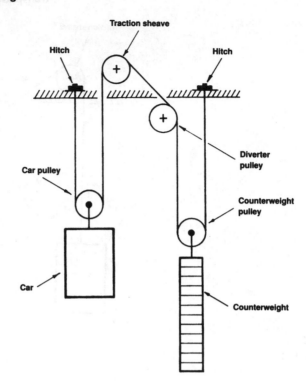

Figure 10.3
Single wrap 2 : 1

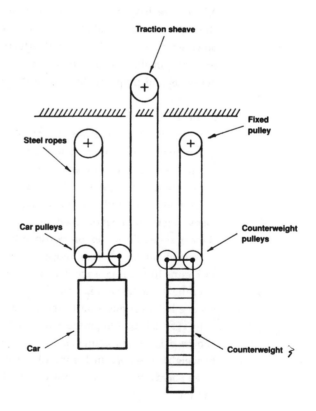

Figure 10.4
Single wrap 3 : 1

Table 10.2 Effect of wrap ratio on car speed

Single wrap ratio	Car speed
1 : 1	x
2 : 1	$x/2$
3 : 1	$x/3$ and so on

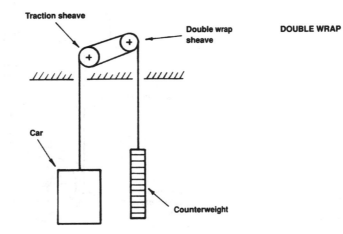

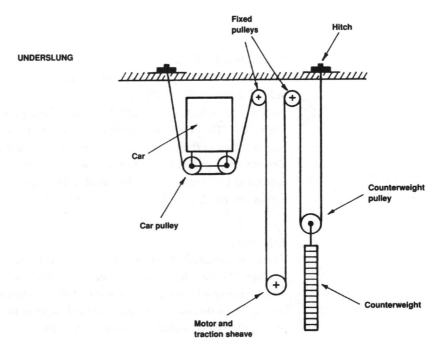

Figure 10.5 Alternative roping arrangements

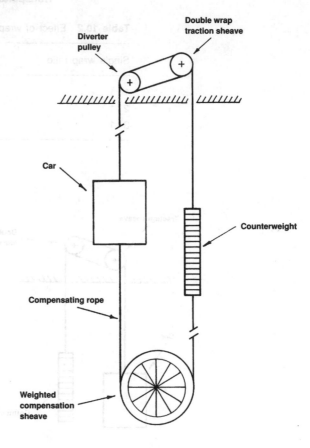

Figure 10.6

Compensating ropes

Labels on figure:

Diverter pulley

Double wrap traction sheave

Car

Counterweight

Compensating rope

Weighted compensation sheave

Emergency braking

The car is retained upright and carried smoothly by guides and channel each side as shown in Figure 10.7. In the unlikely event of rope failure, an overspeed governing mechanism will effect an immediate brake of the type shown in Figure 10.8. The emergency brakes are activated by a continuous rope passing over a pulley in the pit and an overspeed governor pulley in the motor room. The governor locks in response to flyweight inertia from the centrifugal force generated by excess speed, thus jerking the rope in the process. Figure 10.9 shows the position of the governor rope and pulleys, relative to car travel.

Lift doors

These are required in two components, that fitted to the lift car and to the landing. Landing doors must be incombustible, preferably of sheet steel construction over a light steel framework of about 30 mm overall thickness. They usually slide sideways, although vertical movement is used for some industrial applications. Various functions are shown in Figure 10.10.

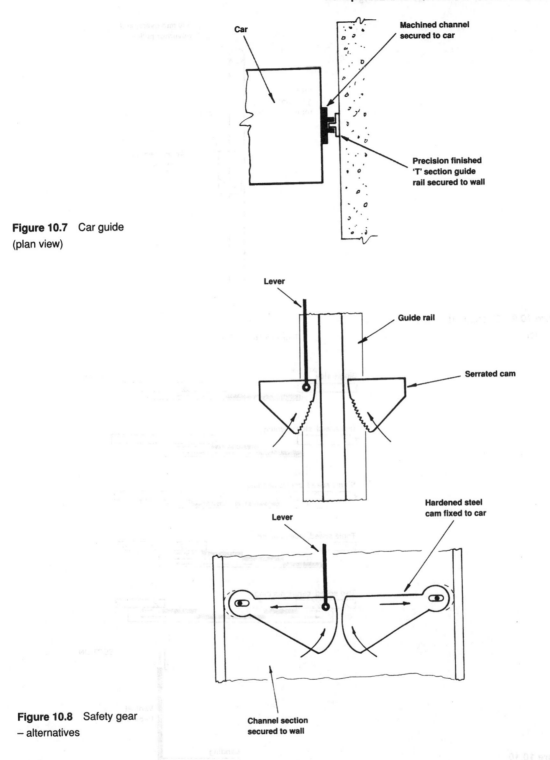

Figure 10.7 Car guide
(plan view)

Figure 10.8 Safety gear
– alternatives

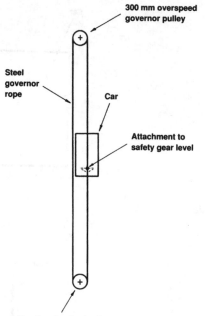

Figure 10.9 Overspeed governor

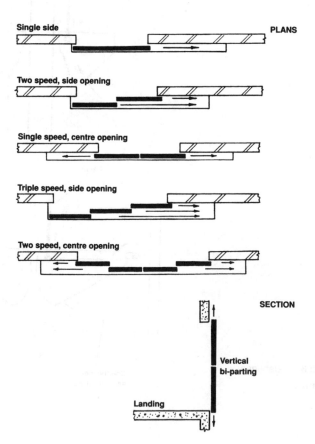

Figure 10.10
Lift doors

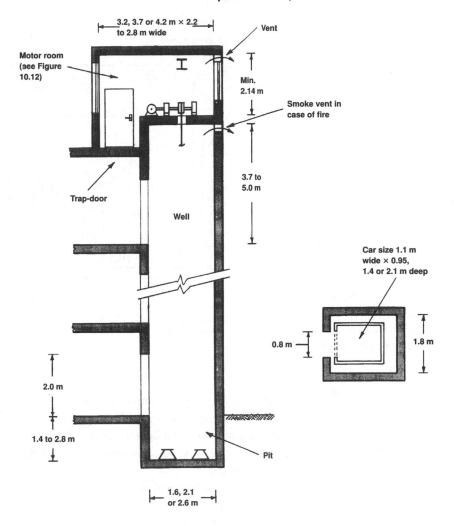

Figure 10.11 Section through typical small car single lift well

Constructional dimensions

Lifts manufactured to individual dimensional specification are possible, but very expensive. Economies of scale dictate standard dimensions to suit all but extreme client or obscure building requirements. BS 5655 provides these standards which have been co-ordinated with manufacturing processes and building applications. Some variables apply, depending on car-carrying capacity, and Figure 10.11 shows typical building and component dimensions for a single lift installation.

Machine room

The machine room is normally located above the well, containing winding gear, traction sheave, control panel, overspeed governor and numerous other components shown in Figure 10.12. Noise from motors and winding gear must be contained with adequate insulation and absorbent bedding for

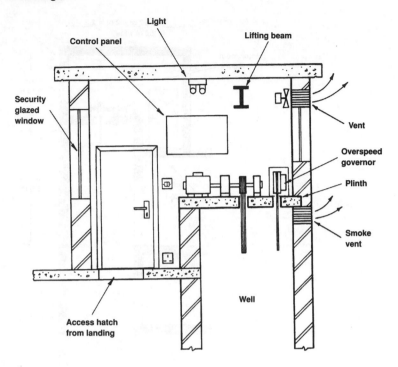

Figure 10.12 Section
through motor room.
Note: Lift motor on
anti-vibration mountings

machinery. An overhead universal beam for raising and lowering equipment
and parts during maintenance is essential, as is adequate daylighting and
supplementary artificial light. Heating will be needed to maintain a minimum
temperature of 10 °C and fan-assisted ventilation to remove excess heat from
electrical plant.

A locked door (key with security staff) provides the only direct access to the
machine room, except for a trap-door directly over the landing area. This is
specifically for raising and lowering items of machinery.

Pit

Below the lowest landing level is the pit, containing buffers. For slower lifts,
spring-type buffers are adequate while higher-speed lifts use oil loaded. Depth
of pit varies from 1.4 to 2.8 m, depending on lift specification.

Brakes

The traction sheave drive shaft is fitted with an electromechanical brake. When
the lift is moving, the electrically operated brakes are lifted clear of the brake
drum, but as the electricity switches off to disengage the motor, spring retainers
activate the brake. In addition to the overspeed governor and buffers, this
provides another safety feature which would activate if the electricity supply
failed.

Shaft

A lift shaft should incorporate the following features:

- watertightness
- means of drainage
- plumb, vertical sides
- smooth painted finish
- ventilation void for emission of smoke
- permanent inspection lights
- have no other services except those necessary for operation of the lift.

Lift controls

Control arrangements include the following possibilities:

- operator
- automatic
- down collective
- directional collective
- group collective
- programmed control.

Operator

Lift operators are rarely employed now, due to cost and advances in automatic controls. They are occasionally engaged in prestige buildings and in hotels for the benefit of special guests. The normally automated controls will have a manual override, to permit callers at each floor level to register in the lift for the operator to respond.

Automatic

This system responds to one call only from either lift car or landing. No further calls are accepted until the car is at rest, therefore it is only really suited to light occupancy and low-rise buildings up to four or five floors.

Down collective

A call button is located at each landing entrance and a set of buttons in the car corresponds to each floor. Landing calls are stored and answered in sequence as the lift car descends. In the upward direction, passengers are distributed in floor sequence by selection within the car.

Directional (up and down) collective

Two call buttons are provided at each intermediate landing, one for up and the other for down. The lowest and highest landings only require one button. A full set of destination buttons are provided in the car. Landing callers simply press the direction button and the call is stored. On a downward journey, the lift stops at all floors where downward callers are waiting or where passengers want to get out. Likewise upwards, operating in sequence in response to stored calls.

Group collective

Where groups or a bank of lifts occur in large buildings, an interconnected collective stored control system can operate. This permits the closest lift travelling in the desired direction to respond, rather than passengers waiting for one specific lift or having to press every lift's call button.

Programmed control

This is an improvement of the group collective system, incorporating time-controlled functions, where demand is known to be particularly high on some floors at certain times. The lift cars can be programmed to be available at the ground floor during arrival times and at the upper floors during departure times. This lends itself to routines found in office blocks, where regular hours are worked.

Hydraulic lifts

Before electricity became widely available, some of the earliest lifts were operated by hydraulic water power. Later experiments proved oil to be a more efficient medium, but with an overall theoretical maximum travel of 21 m, they are no threat to electric lifts for higher-rise buildings. Nevertheless, there has been an enormous growth in the demand for oil hydraulic lifts and they now constitute most of the lower-rise buildings market sector.

Hydraulic lifts have several advantages:

- capacity for very heavy loads
- accuracy in floor levelling
- smooth ride characteristics
- low-level plant room
- no structural loads from winding gear
- pump room can be located up to 10 m from the shaft.

However, they are limited in speed to about 0.75 m/s to maintain adequate standards of control and comfort. Depending on the system used, specialist equipment may be needed during construction, to provide a deep borehole to accommodate the hydraulic cylinder. Figure 10.13 shows the principle of operation, Figure 10.14 the application and Figure 10.15 some variations.

Fire-fighting lifts

The concept of a fireman's lift developed during the 1920s and 1930s, as the need for rapid emergency access responded to the growth in numbers of tall buildings. The original concept was a variation within a conventional passenger lift, which contained a priority break-glass key switch. This was normally at the ground floor, and when activated it brought the lift to that floor immediately.

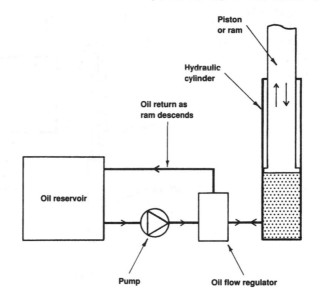

Figure 10.13 Oil
hydraulic lift – principles

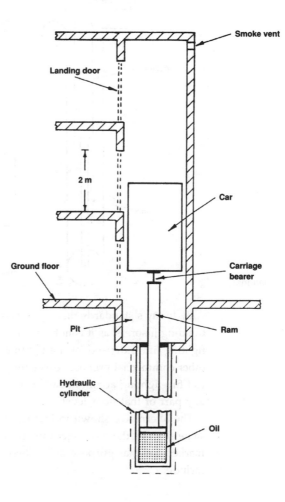

Figure 10.14 Oil
hydraulic lift – application

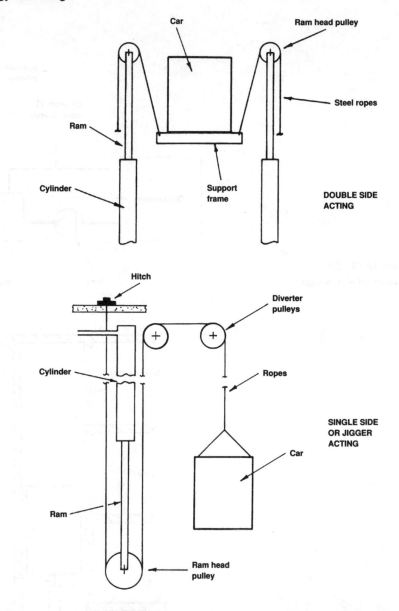

Figure 10.15 Oil hydraulic lift – variations

By today's standards this is unlikely to be acceptable to the fire service or building insurers, as it not considered adequate as a purpose-designed fire-fighting lift. Independent fire-fighting lifts are required in offices, shops and other commercial premises exceeding 18 m in height. Figure 10.16 shows typical fire-fighting accommodation in a shaft located no more than 60 m from any part of that floor level.

The alternative shown in Figure 10.17 has the fire-fighting lift sharing the same shaft as the passenger lifts. In this situation the fire lift must be clearly marked for that purpose only. These lifts will require specific provision, to include:

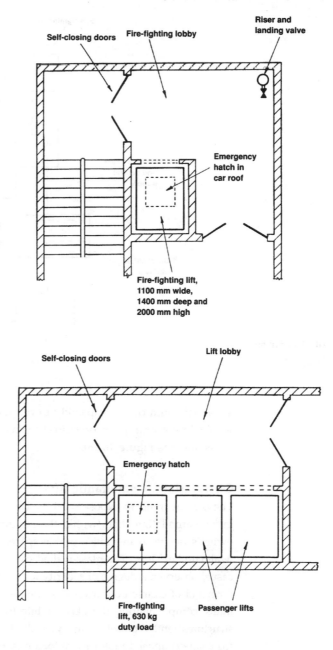

Figure 10.16 Fire-fighting lift – purpose made

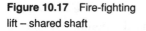

Figure 10.17 Fire-fighting lift – shared shaft

- 630 kg minimum duty load to accommodate fire-fighting equipment
- minimum internal dimensions of 1100 mm width, 1400 mm depth and 2000 mm height
- an emergency hatch in the car roof
- manufactured from non-combustible material
- a two-way intercom
- 1 hour fire-resisting doors of 800 mm minimum width × 2 m height

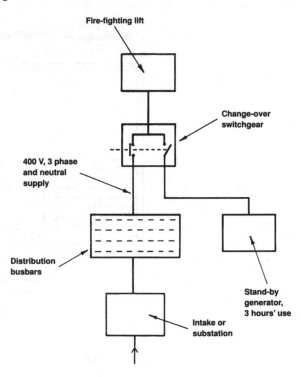

Figure 10.18   Fire-fighting
lift – control diagram

- a maximum of 60 s capability to run the full building height
- dual power supplies, one direct mains and the other an emergency generator (see Figure 10.18).

Observation or panoramic lifts

The current trend to expose services, largely due to the increasing difficulty in accommodating them within the building fabric, as well as the building owner's desire to maximise usable space, has led to some very innovative 'wall climber' lifts. The glass-walled cars provide a focus of interest for the casual observer, a degree of security for occupants, a mobile observation platform and of course floor access for the user. They are very popular in atrium malls, complementing the glass architecture. These modest rise lightweight structures lend themselves to hydraulic lifts, freeing the building designers from superimposed motor room loadings. Figure 10.19 shows possible modification to a standard car and shaft, to provide visibility on one side. Glass doors and sides could also be provided for all-round visibility.

Paternoster

Paternoster lifts are very common in America, but have not been so popular in the UK. They consist of a series of doorless cars usually of two-person capacity, suspended on continuously moving endless chains. Figure 10.20 shows

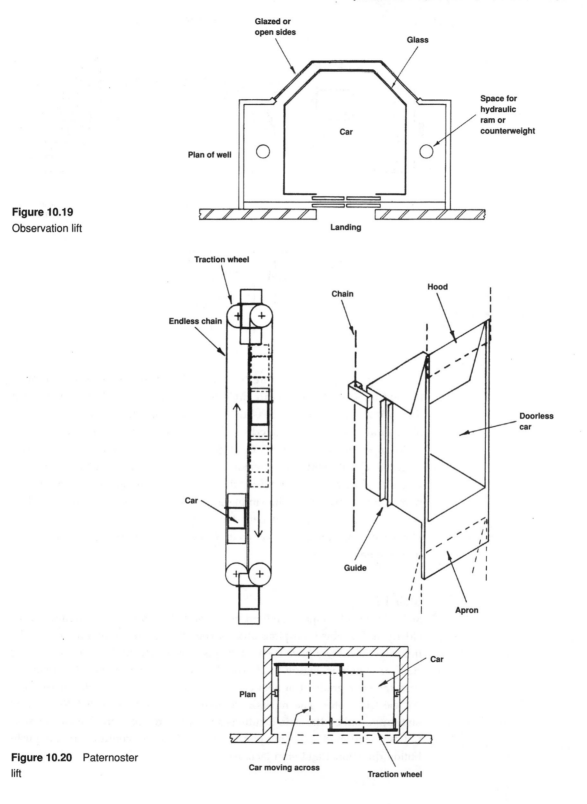

Figure 10.19
Observation lift

Figure 10.20 Paternoster lift

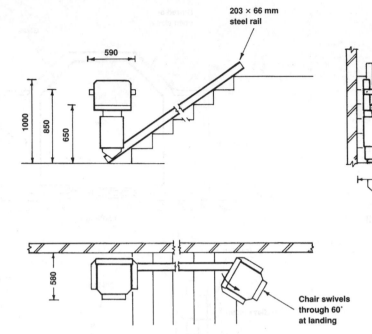

203 × 66 mm
steel rail

590

1000
850
650

Chair folds
to 320 mm

580

Chair swivels
through 60°
at landing

Figure 10.21 Stair lift –
approximate dimensions
(mm)

the concept, which in the interests of safety is limited to a speed of no more than 0.4 m/s. Safety devices are fitted to the leading edges to halt movement if a car is misused. Nevertheless, they are not suitable in public buildings and other locations where the elderly and infirm are likely to gain access.

Paternosters are most suited to single occupancy buildings such as offices, where familiarity with the system and a high degree of staff mobility is a feature. For efficiency and simplicity they have many advantages over conventional lifts. In particular the simple geared constant speed motor, which is subject to less wear than the persistent stopping, starting and speed changes of normal lifts. Landing and car controls are also unnecessary, with the exception of emergency buttons.

Stair lift

Stair lifts have become a standard means of vertical transport in homes for the elderly and disabled, hospitals and conventional homes containing physically infirm people. They have been developed for simple application to domestic stairs as shown in Figure 10.21. The chair moves up an inclined rail parallel with the stair gradient at about 0.15 m/s, powered by a single phase 230 V AC electric motor. The rail is a standard steel joist bracketed to the wall and supported by the stair. Landings and turns are not a problem, as the steel rail can be formed accordingly. Transformed 24 V DC controls provide push-button directional and stop facilities.

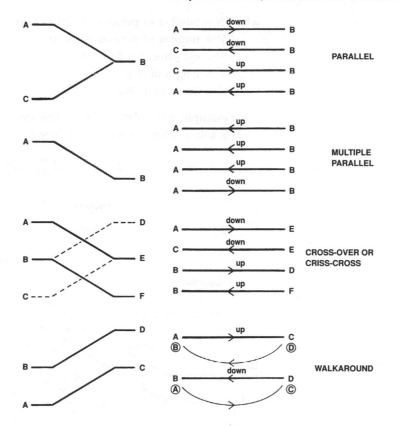

Figure 10.22 Escalator arrangements

Escalators Escalators are moving stairs designed to provide efficient vertical conveyance of people. In low- to medium-rise buildings they will compete favourably with lifts, even though they only move at between 0.5 and 0.65 m/s. In high-rise buildings the space will not be justified and the modern high-speed lift provides a superior service. Where large numbers of people are anticipated, such as airports and railway terminals, department stores and shopping malls, several escalators will be required and can be grouped in a number of ways to suit the building function, as shown in Figure 10.22.

The angle of inclination is normally 30°, but may be increased to 35° if the vertical rise does not exceed 6 m and the speed is limited to 0.5 m/s. Figure 10.23 shows typical dimensions for an escalator, where step widths are standardised at 600, 800 and 1000 mm. Step or tread length is a minimum of 400 mm.

Escalator capacity

The following formula can be used to ascertain capacity and compare efficiencies and suitability of escalators at building design stage:

$$N = \frac{3600 \times P \times V \times \cos \theta}{L}$$

where N = number of persons moved per hour

P = number of persons per step

V = escalator speed (m/s)

L = length of step (m)

θ = angle of incline.

For example, an escalator of 30° incline, one passenger per step, a speed of 0.5 m/s and a 400 mm tread or step length:

$$N = \frac{3600 \times 1 \times 0.5 \times \cos 30°}{0.4}$$

$$= 4500 \times \cos 30°$$

$$= 3897$$

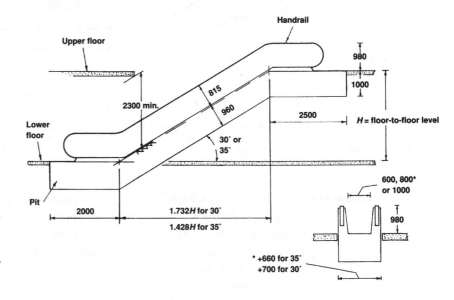

Figure 10.23 Escalator dimensions (mm)

Spread of fire

The void containing escalators could encourage fire to spread rapidly through a building. The following precautions could be considered and may be an insurance requirement:

- sprinklers, installed to provide a continuous curtain of water down the escalator void
- fire curtains or shutter mechanism released by fusible link or smoke relay to seal the top of the escalator shaft
- compartmentalisation or separation of escalators into a well or fire-protected enclosure.

Travelators These are otherwise known as autowalks, passenger conveyors or moving pavements. They provide horizontal movement of pedestrians, wheelchairs, luggage trolleys and small vehicles up to the practical limitation of about 300 m distance. They are particularly useful in large railway and airport terminals and may be inclined up to about 15° where level differentials occur. Speeds range between 0.6 and 1.3 m/s, limitations being imposed because of the difficulty in getting on and off. Combined with walking, the overall pace could be about 2.5 m/s.

Materials for travelators must be flexible or elastic and include reinforced rubber or composites and interlaced steel plates or trellised steel. The latter two have the facility to deviate from the conventional straight line. Developments have been proposed for variable speed lanes, where it is possible to transfer sideways from say a 1 m/s conveyor to 3 m/s and then 5 m/s, transferring back as the end appears. However, if bunching were to occur the result could be disastrous.

Index